MORE WAYS THAN ONE OF FIGHTING A WAR

Eric C. Claxton, O.B.E., B.Sc., F.I.C.E.
Founder and Hon. Organiser of Casualties Union

With grateful acknowledgements to the artist, Mr. Arnold Beauvais

MORE WAYS THAN ONE OF FIGHTING A WAR

The Story of a Battle School for Civil Defence

and

The Creation of Casualties Union

Remembered by

Eric Charles Claxton

The Book Guild Ltd
Sussex, England

The Book Guild Limited
Temple House
25 High Street
Lewes, Sussex

First published 1990

Set in Baskerville

Typesetting by Book Economy Services
Burgess Hill

Printed in Great Britain by
Antony Rowe Ltd
Chippenham

British Library Cataloguing in Publication Data
Claxton, Eric
More ways than one of fighting a war.
1. Great Britain. Medical services, 1939-1945. Biographies
I. Title
940.54'7541'0941

ISBN 0 86332 431 2

ACKNOWLEDGEMENTS

I ACKNOWLEDGE most gratefully the assistance given me during the preparation of these notes by my wife Helen, John Hepworth, Marge Clouts and John Stevens with their sympathetic reading of my drafts and I hope that I have been able to take positive advantage from their comments and suggestions. I acknowledge with gratitude the work and support of all those mentioned in the text who strove so courageously in those difficult days; Jarché and *Illustrated* for the photographs published in that weekly paper; Richard and Moreen Beel for reproducing clear pictures from well-worn, rather tatty and somewhat dog-eared prints; Arnold Beauvais for his impression of myself, drawn some forty years ago from a wartime photo at the request of John Stevens; the Surrey County Council and the Home Office for their goodwill in the publication of these notes and my warmest thanks to Brenda Whiteley for contributing a Foreword.

Rescuemen attend to lacerated back – difficult to bandage firmly.

IN DEDICATING this book to all those, especially the 'casualties', who endured the rigours of the rescue schools in Leatherhead and Eindhoven, Holland, I pay homage to their courage and their devotion to the well-being of their fellow men and I thank them all from the bottom of my heart. They understood the difficulties that had to be overcome and they joined me in saying, 'I CAN AND I WILL', I admire them all.

Their spirit may perhaps be summed up best in the well known prayer –

> 'Teach us, O God, to give and not to count the cost, to fight and not to heed the wounds, to toil and not to seek for rest, to labour and not to seek any reward save knowing that we do Thy will.' – *Acts of Devotion.*

Contents

LIST OF ILLUSTRATIONS

FOREWORD

I AM GRATEFUL for the opportunity to welcome the writing of this account of the creative activities undertaken at the bombed site of St Andrew's Convent, Leatherhead, by the Surrey County Rescue School, between 1941 and 1945. The challenge to rescue one's fellows from bomb damage was ever present and great, and the need for realistic training and research became of utmost importance.

The school staff attempted to face up to the challenge in a realistic and urgent fashion and gave training to civilian men and women in ARP and Civil Defence and also to members of the Allied Armies. What they needed were 'casualties', who were prepared to stand in during training and rehearsals for the real casualties a few minutes, hours, weeks, or months later.

I was happy and proud to play a part in this training and I pay tribute to the tireless efforts of the staff of the school and members of Casualties Union who maintained a high degree of enthusiasm over those hectic years. I hope that this book will be widely read and that it may encourage others to work on similar lines so as to care for those injured in the peace-time world.

Brenda O. Whiteley
Founder Vice-President Casualties Union

PREFACE

At the time of writing these notes many years have elapsed since the days of war and battle training – thankfully long past. I have often wondered why this heavy responsibility was given to me? And what, in fact, I was being asked to undertake? Sometimes, I think that far from having that responsibility set upon me, recognising what needed to be done, I created my own opportunities; but of one thing there is no doubt, the enemy created the problems and the necessity for the work, which I found myself called upon to undertake. It was a tremendous challenge, which I embraced with enthusiasm. I had to find the inspiration, the drive, the means and the methods to confront disaster head-on. My previous life served, quite simply, as preparation for this battle against bomb damage and injury.

When I was five-years-old the First World War broke out. In 1920 I learnt first aid as a Boy Scout. Throughout schooldays I learnt something of leadership in class, on the sportsfield, in the OTC and as a prefect. At university leadership training continued, and I learnt the field skills of the sapper in the senior OTC engineers. My professor urged me to learn first aid, and I took my St John Ambulance first aid certificate through a class at college in 1929. My need to know first aid became apparent in my second job, when assistant clerk of works on the construction of the West Middlesex Main Drainage Scheme, with its disposal works at Mogden. As a Scouter, I taught first aid to both Scouts and Cubs. In the process I learnt a lot about boys and men.

Gaining Certificate 'A' infantry in the junior OTC in 1926 and Certificate 'B' engineers in 1929, apart from my degree in engineering in 1930, and corporate membership of the Institution of Civil Engineers in 1934, I was entitled to a commission in the army. Indeed in 1932, while unemployed after being a victim of Geddes' axe, I considered the possibilities of making a career in the army, but, by then I was already hooked on bridge building, after a year examining bridge designs in schemes submitted for grant purposes to the Ministry of Transport, HQ Bridges Section. And I was greatly excited by the new Sydney Harbour bridge, and the magnificent bridge built some seventy-five years earlier over the Firth of Forth, and decided to remain a civilian.

In 1939 I had been working in the Highways and Bridges Department of the Surrey County Council for four years, and all my work had been to do with bridges, and the ancillary road, rail, river and drainage work. When the threat of war arose, in common with all holders of Certificates 'A' or 'B, I informed the War Office of my whereabouts and occupation, requesting their instructions. The County Council listed me among their reserved staff and when the War Office replied, it was to instruct me to stay put. At least twice the Highways and Bridges Department of the Surrey County Council formed and staffed companies of Royal Engineers. Each time I applied for a commission, only to be told once more, 'You are reserved'.

The Surrey County Council moved over to a state of war readiness during the latter part of August 1939. The whole design and drawing office staff in the County Engineer's Department set aside its normal work and prepared maps showing the boundaries of the various district authorities, the areas served by the different gas companies, electricity undertakings, water companies, telephone areas and drainage areas. These were for use by emergency control centres in both Metropolitan and ex-Metropolitan areas of the county. The staff were divided into two divisions, one to serve each area. I was instructed to report to Guildford by Sunday, 3 September.

On that day, we were expecting a broadcast by the Prime Minister, Neville Chamberlain, at 11 o'clock. I arrived at County Hall to collect my office instruments just before that time and listened to the broadcast there. No sooner had the PM announced that we were at war with Germany, than the siren sounded the alert. My colleagues took to the shelters, while I dashed off to Guildford. Fortunately it was a false alarm and I arrived safely at our new headquarters in a nearly completed school building in Stoke Park, Guildford. We lived there with candles for lighting, because as yet there was no blackout provision.

The first major change of my life was that, whereas for four years I had travelled the eighteen miles from home to County Hall and back by pedal cycle, from the outbreak of war onwards all my journeys were made by car.

My first assignment in this new world was to measure and make a register of every bridge in the whole county; its precise location, size, type and form of construction; so that, in the event of it being bombed, suitable materials could be sent out immediately to make repairs. There were some nine thousand bridges in the combined Met and ex-Met divisions of Surrey. The schedules would be a great help to engineers on duty at ARP County Group control centres at Kingston-on-Thames and Dorking.

My second task was to wind up a number of construction jobs, so as to take full advantage of all work completed so far, but not to continue them further.

My third task was shift engineer at Dorking ARP control on a rotating eight hour shift. I was permitted to live at my home in Sutton, because it was nearer to Dorking than Guildford and in spite of the fact that my home was in the Met Group area. In due course I was instructed to report back to Guildford and all of a sudden my life was changed once more; this time permanently. I was launched on my mission to train rescue men and in the process to create Casualties Union.

Casualties Union is now forty-five years old and has spread to many communities across the world. The Union continues to mount exercises to give all emergency aid workers, who desire it, extra and growing experience, especially in handling casualties in awkward situations and essential practice in diagnosis, without which emergency aid might be fruitless.

I hope that this book may whet the appetities of many adventurous young people to make emergency aid a major plank in their life's platform. It may perhaps stir elder people to polish up and use techniques that they dabbled in earlier in their lives. I wish them all well and offer to them the open hands and hearts of Casualties Union.

Eric Claxton

1

Launching the Rescue Schools

The County Engineer, W P Robinson, C.B.E., sent for me. He explained that it was a matter of urgency; he was forming two rescue schools in our division, one at Redhill and the other in Guildford. Both were in Borough Council yards. He had appointed two instructors, S H Thomas and J Bullen, who were already in training at a rescue school at Kingston-on-Thames. He added, however, that he needed equipment for the schools and handed me a list of tackle not readily available in shops, adding also this further comment, 'A man of drive could equip both schools in forty-eight hours'.

I was not to know that he was appointing me godfather to the two schools and also to seven County rescue parties that he was to recruit a little later from the County's five rural divisions, but I accepted the challenge by commissioning here, begging there, borrowing somewhere else and buying wherever possible and had the schools ready to start in twenty-four hours.

It took a little longer to get hold of some equipment to replace what I had cadged on loan, like heavy jacks – the seven County rescue parties would need similar equipment too. I was also to arrange for District Council's rescue parties to attend for training, so that the schools were fully occupied.

Both rescue schools had already been furnished with a scaffold tower some 12ft square and 30ft high from which rescue parties could practise lowering one another and at which they could learn to use extending ladders, tackle and ropes generally. Perhaps of most importance I found both of the instructors were strong characters with a lifetime of experience of building and demolition, but when it came to dealing with rescue parties coming for training from eighteen local authorities, several with more than one party, and many of whom were volunteers, it seemed that the instructors also needed a godfather of sorts. To be anything like that I had to learn what the instructors taught; what they had themselves been taught, as well as to discover what rescue work was really about. I went to the Kingston school as well as the schools at Redhill and Guildford. I attended a demonstration at the Crystal Palace, where a contest between various London units lowered people from the top of a similar tower to those built at Redhill and Guildford.

At that time there was no bombing, but one was concerned that the person being lowered, when real trouble came, was liable to be injured. Rescue parties, I was told, would be given first aid training by the voluntary aid societies and in our area St John or Red Cross undertook the duty. My task was to ensure that the men were familiar with all their equipment and able to get somebody safely from a height, out of a hole, through a window opening or up or down stairs – something of a mechanical problem of handling a load of between one and two hundredweight and of course with care, because the load was a human being. In normal training one or another of the rescuemen was told off to be the "load". There was often a bit of ribald humour concerning the victim, but he was always well protected by his working clothes and blankets. I soon realised that the best way to put each team on its mettle was to offer myself as "casualty" at certain stages of their training and I tried to be present at the right moments with each course.

My civvies were much lighter than the clothing the men wore. As yet we had no uniform and although I was athletically fit, I was probably not so hard or tough as they were. I was, therefore, a little more sensitive to the handling I received and the discomforts of the methods used. Rope slings could be like reasonably comfortable chairs, or like string loops one almost slipped through – one could be sitting or standing or laid flat on a stretcher. Someone had devised an ingenious method of securing a person to a stretcher with a forty foot long one inch lashing. The feet were well secured and there were thongs across the body below the knees, hips and waist and the whole made totally secure by thongs across each shoulder. There was no support for the head and the comfort in being secured to a standard issue metal ARP stretcher depended almost entirely on the protection afforded by two blankets that were included in the method. So as long as the method of lifting or lowering kept the stretcher in a horizontal position and one was able to see the nature and correctness of the knots being used, one could relax with a fair degree of confidence. However, when one was upended to pass through a small hole in the floor like a trap door, one might slip precariously within the lashing; or worse still, when firmly tied and upended over a window sill to be lowered clear of the building, there was a moment while being launched when the window ledge was under the knees, that the stretcher flexed slightly, tightening the thong across the shins in excruciating pain – momentarily only if the team knew their job. Thus I began to be conscious of the sensitive spots in the uninjured body, some of which are not at all obvious to beginners. My reader will realise that I learnt many of these points the hard way.

After only a few weeks it became inconvenient for the Reigate Corporation to continue to give space at their Redhill depot for a rescue school, because the space was needed for an emergency food store. A place was found for us to transfer to at Wonham Mill, Betchworth, near Dorking. This was in many ways an ideal site, but it had been used over the previous few years for poultry farming and although it was reasonably well mucked out before our occupation, parts were still in use for hens and those that had been removed had not taken all their parasites with them. These creatures were quite happy to make their temporary home with any human being who stayed around a while. The mill had high windows and there was an outside wall to the yard, so instructor Thomas liked teams to land their casualty outside that wall. At certain stages while being lowered on a three inch rope and steadied by two forty foot lashings as guide ropes, one felt as though one was bouncing on a tug-of-war rope and it was with great relief that finally one landed safely.

Everytime a bomb dropped anywhere near our area, I went to see what the wreckage was like. It became obvious that in the bulk of the buildings in Britain, built with brick walls and wooden floors and roofs, the wreckage from bomb-blast left piles of bricks, mortar and furniture sandwiched between the platforms of floors and roof, often leaving the stairs and chimney stacks still standing. It seemed obvious to me that we

Bomb debris showing cavities.

Heap of debris.

needed a different type of training ground for our schools where we could experiment on ways of getting people out from the cavities left in the sandwiches, without the piles of bricks, mortar and plaster collapsing further and suffocating or crushing the victims still more.

In the spring of 1941 I voiced these thoughts to the County ARP Training Officer and it seemed only a matter of days later that he phoned me the news that St Andrew's Convent, Leatherhead, had been badly damaged by a parachute mine the night before and had been evacuated by the nuns, who had moved to an old house nearby. I went straight away to commiserate with the Mother Superior and to ask her if I might use the site for a rescue school. Her reply was as immediate as it was straight forward, 'Yes', she said, 'thank God some good may come out of this disaster'. Thereafter, almost every week the dear lady visited me at the rescue school to enquire how we were faring.

2

A New Beginning

To transform the wreckage into a rescue school demanded a lot of thought and a lot of hard work. Lecture room, canteen facilities and sanitation had to be devised for the students and some accommodation for the school staff. Between weather-proofing some of the bombed rooms that were reasonably intact and importing wooden huts, primitive accommodation was provided for everybody.

The craftsmen in the County Rescue Parties were augmented from time to time when additional skills were required, thanks to the cooperation of the County Divisional Surveyors. The school had then to set about getting rations of such items as soap, which by the nature of our work were always in great demand. However, the most difficult task was to make the wreckage into graded areas where different problems of rescue could be posed and where a full course of exercises and investigations could be carried out week by week. Thus a pile of five hundred tons of debris was gathered together with articles of furniture and sections of unbroken floor, roof and wall incorporated to create cavities within – not just a heap of brick rubble. There were places where tunnelling could be undertaken to give access through brick rubble to cavities beyond.

Two floors, both collapsed and sloping, one on top of the other, were developed to study lifting somebody out from a 'vee', where gravity tended to wedge the victim in amongst the debris at the root of the 'vee', especially while attempts were being made to move the person. Places were made where people could be lifted out of holes or lowered through. There were larger holes in some floors where persons on stretchers could be brought through and also holes through walls – all requiring the greatest care for the victim being handled.

It was necessary to have a variety of loads of different

materials, sizes and weights – blocks of stone or concrete, steel girders, baulks of timber, tree trunks – any of which could have trapped a human being. The men needed to learn how to estimate the weight involved so they could select the best equipment that was ready to hand or to know when to call for additional assistance. It was also necessary for them to know how to deal with flowing water, or fire, or gas attack. Suitable places were incorporated, as well as places for the exercises undertaken at the former schools to be continued.

Such a recital indicates the comprehensive nature of the new school, which replaced both the former schools, but it also indicates the courage, patience and self-control demanded of both 'victim' and rescuers going to his aid. For most of us there was a serious risk of claustrophobia in the small spaces beneath the debris and/or vertigo in climbing to high places along narrow walls.

To combat this the course started with a climbing frame, long pipes and a debris crawl to convince the students how small they were in reality and what could be done in such small spaces. They walked along a line on the floor; then on a four and a half inch wide wall a few feet high and finally along a similar wall thirty foot above the ground. We found that with these exercises we could give students the confidence to cope, but they also needed the physical resilience to endure. It so happened that at that time there was a radio programme early each morning called, 'Up in the morning early', which brought rhythmical exercises to music and which I had found helped me to keep fit enough to cope with my end of this task. I chose a series of exercises which seemed appropriate, and to the students' horror, we started each day with my version of, 'Up in the morning early'.

Much of our work involved the use of ropes. There were two classes of rope in the standard issue – three inch and one inch. The dimensions refer to the circumference of each rope, so the thicknesses were approximately one inch and three eighths of an inch respectively. The heavier was used with pulleys and tackle generally and was about one hundred feet long; the lighter ropes were forty feet long for lashings, guide lines, slings and safety ropes. These were all used in an endless variety of situation and the men had to learn how to bend the ropes or sheets – to tie secure knots and such knots must always be secure for lifting wounded people or, indeed, heavy loads.

Some of the men were right-handed and others left; some found it best to copy someone else, but as one watches someone else with similar hand, that person appears to be opposite handed – a bit like looking at oneself in a mirror. It was convenient, therefore, if the instructor was ambidextrous, so the students could follow each movement as the instructor bent the rope on the opposite hand. One could, of course, stand behind the student and reach forward to take hold of his hands and guide his movements, so long as he was not too stout. Later on, when women came as students, this method became more popular.

Most rescue operations began at night and, because lights would have aided the enemy to press the attack, it was necessary to work as much as possible in darkness. This proved very difficult for most people because one tends to use one's eyes to get knots correct. The final stage, therefore, in the use of ropes and lashings was carried out in total darkness. It was only in the dark that men learnt truly to use their hands and the skill with ropes was mastered. Subsequently, and indeed much later, we discovered that this was also the best way of mastering bandaging. The quality of work improved, the tension became uniform, the bandage firmer and the tendency for the dressing to slip during the binding of the injured part was eliminated, leaving the bandages more effective and comfortable. The skill of the hands was only fully realised when they were no longer dominated by the eyes.

Methods of reaching wounded persons were studied and means of lifting them clear of the wreckage were practised until each group of men could lift a man on a stretcher from a wide variety of places, or without a stretcher depending on the nature of the injuries and the physical possibilities of the situation. The exercises included removal from an upper floor, through an upstairs window in horizontal, or vertical, position, up or down stairs, up or down ladders, up or down through a trap doorway or manhole, coalshoot or other opening in the floor or ground. If the hole was too small for a standard stretcher to be used, a narrower stretcher had to be improvised, (such as a cupboard door, a short ladder or a chair) or else the casualty had to be lifted in a sling of appro-

Casualty on stretcher being landed well clear of damaged building by telpher line

Unconscious man fireman's lifted from gas-filled area using remote breathing apparatus

priate design for the situation. This depended on the size of the person and the size and shape of the hole, as well as the sex of the casualty and the nature of the injuries. Slings may be formed from a rope, (a bowline, bosun's chair or a double bowline) a canvas seat or a Holger Nielsen stretcher.

The choice in every case needed to take into account the nature of the injuries as well as the physical situation. If legs were broken, it was impossible to use a sling round the ankles, otherwise the quickest way of lifting someone out from a manhole because the body is at its narrowest; if the ribs are broken it is equally impossible to use a sling round the chest. In cases where it was suitable to use a sling round the chest a larger space was required as the arms must be held by the sides, the shoulders remaining square -perhaps being able to use the diagonal position through a square hole. If then it was necessary to raise the arms above the head in order to narrow the shoudlers, a sling under the crotch, or similar, becomes essential, when the wrists can be tied to the lifting ropes above the head.

While these many methods could be practised as drills or exercises with one of the students acting as the patient, little gentleness could be expected from the robust men in thick clothing with one of their colleagues. At the next stage, the students could learn to select the most appropriate method of handling and movement, when they were told the nature of the injuries that the victim was supposed to have sustained. Only when the students were able to examine the casualty themselves to determine the nature of the injuries and so become aware of the nature of the pain, weakness and loss of functions, was complete consideration, tenderness, sympathetic handling and true reassurance achieved, and the students were able to graduate.

This was not achieved until we had enrolled able volunteers who would enact the complete behaviour of the injured, distressed and agonised casualty. However this was to come about in an interesting and somewhat dramatic way later on.

The courses went well and the scope was extended as the rescue men gained in skill until the County men had all been through a series of courses. But things were not always straight foward. On 14 February, 1942, a fire occurred during the night, burning out my office and the instructors' room above. My cine camera with which I had started to make training films was destroyed as well as records and other important items.

3

Research and Development

It was no sinecure to find oneself responsible for the creation of a rescue school handling all the problems liable to be faced by rescue workers in the field of bomb damage, with fire, drowning, asphyxia added to the hazards of damage and trapping amongst debris. To make it worse nobody seemed to know much about rescue from bomb damage, because the sum total of experience appeared to be limited to the Spanish civil war and the scope of first aid textbooks. Everything had to be researched and worked out amidst the debris. Add to this the restrictions of rationing of food and clothing for men and women in a business that was both fraying and exhausting. The one concession that a rescue man got was a pint of milk, while the normal civilian ration was only half a pint. The challenge was most searching and, properly handled, brought out the best from each one of us.

To start with I got a lot of, 'I can't do that, guv'nor', so since it just had to be done, whatever it was, I wrote, 'CAN'T', in large letters on the lecture room wall and crossed it out in heavy red paint and substituted, 'I CAN AND I WILL'. I had a poster made and put it up in all the depots. I never heard, 'can't', again and, 'I can and I will', has remained my motto ever since.

However I concede that the whole business was terrifying and, because I was convinced that the way to save life was to exploit the cavities in the wreckage, where trapped casualties had the best chance of survival, it was essential that we learnt to make the most of these cavities, without dislodging the heavy beams, floors, girders, bricks, tiles and plaster dust, any of which could kill or maim those already wounded. I was fearful too that the very natural temptation to climb over the wreckage might crush, smother or impale those trapped underneath. We had to learn to go in and through – a terrify ing prospect. Even as one contemplates it, one feels the terror of claustrophobia pressing round one's heart. It was perhaps the worst enemy.

I conceived the idea that the way to overcome this was to prove to each person how small he, and every other person, really was. It was in a sense experimental. The frame that I got constructed was adjustable to allow for the bigger person – they came up to twenty stone, but I never had to use the larger spaces. There were rectangular spaces, triangles, some chest high, some low down, and a full height vertical slot for squeezing through. We discovered that if one could get one's head through wearing a standard steel helmet, the rest of the body would go through also.

The students passed through each of these slots, then wriggled through an eighteen inch diameter pipe and crawled through a thirty inch pipe. Next they wriggled through the eighteen inch pipe towing another person, who was also within the pipe. After that the students crawled through the debris of a living room which, while covered over by debris, had no weight on top. It was not a difficult step from this obviously safe wreckage to something not dissimilar with tons of debris on top of the wreckage to be crawled through. Later exercises in tunnelling, with inevitable minor collapses, taught us to make secure as we progressed. Later on we were to learn that it took a great deal of courage to be a 'casualty' in one of those cavities, while someone blundered through the wreckage dislodging a brick here, and a cloud of plaster there.

The course moved on to becoming knot perfect so that ropes and tackle could be used to extricate a human being from all sorts of apparently inaccessible places. The men became highly skilled at securing a person to a stretcher so that it could be turned on end – necessary for getting out from an air-raid shelter – for passage through a trap doorway, or for security when the stretcher turned over unexpectedly. The persons used as casualties in the early training were themselves or myself. I pretended a nonchalance I did not always possess that their knots were properly formed when they hoisted me out from a third storey window opening! It made them more careful and gave them confidence to feel that I had faith in them.

Climbing frame for anti-claustrophobia training in use in the snow.

Hauling a colleague through an 18" pipe and crawling carrying a colleague through a 30".

I soon discovered that the men were shy in handling other than their immediate fellows. On some midnight exercises, when no moon was to be seen, I introduced them to some females as victims of disaster. I am certain that the men blushed more than the brave volunteers, recruited from my colleagues at County Headquarters. This was only the tip of the iceberg. What were they going to do when women were injured and needed bandaging to stop bleeding or broken limbs made secure. I asked all sorts of questions and discovered that the first aid courses they received at their home depots were woefully inadequate. I appealed to my colleagues in the County Medical Officer's department. Assistant County Medical Officer of Health, Dr W J Maelor Evans, agreed to give a weekly lecture on first aid in wartime and this bridged the gap in some measure. They would know what to do and how to do it, if there was no debris about and they knew what was wrong with the injured person! I was still concerned when circumstances played straight into my hands. The enemy stopped bombing the UK.

The lull tempted district engineers to utilise their rescue parties to carry out urgent work, set aside during the period of constant alert, and urgent rescue training. To me the lull was the most generous opportunity to incorporate many of the essential skills we would be needing in the onslaught that the enemy was preparing for us. I appealed to my chief, County Engineer, W P Robinson, C.B.E., to permit me to run a special course for the Engineer Heads of the Rescue Service, so that I could 'put the fear of God into them'. He responded in fulsome manner. 'Quite right, my boy! I will put my name at the head of the list; then the others will have to come'. He was as good as his word, and they all came for three days (6-8 July, 1942). What an opportunity! What a responsibility!

As I prepared, my chief made it abundantly clear that neither he nor his colleagues would be given instruction by other than professional men – so it was up to me and my doctor. Dr Evans was a great help. I needed casualties that bled, were frightened and who were in pain. He introduced me to members of Byfleet Ambulance Station who dabbled in the make up of wounds and who, under the influence of a certain lady driver, introduced some acting into their own performance for their own training. The lady was on leave! The others came.

The course duly assembled on the 6 July, and I addressed them as I had all the previous courses on the theme L - I - F - E to preserve which the rescue service existed. It went –

L – look see – a need truly to observe.
I – invade the wreckage – a well disciplined but determined invasion.
F – first aid as necessary, like the plumber, to stop the leaks from which LIFE flows out.
E – extricate to safety and second aid in hospital.

This was a fighting talk for men about to go into battle. Then they were let loose on the five hundred ton pile of debris to continue a tunnel that was already begun. After some thirty minutes they came across an injured man, unconscious with a head wound. Bleeding had been profuse, but was beginning to dry. They found the close quarters difficult, once they found the casualty, and I sensed our old friend claustrophobia.

Complimenting them on a gallant attempt and sympathising with them on the tight working space, I quickly turned their attention to claustrophobia and the exercises we were to undertake. I had promised my chief that I would demonstrate myself, before calling upon them to perform. I demonstrated on the frame – quickly dubbed the 'stocks' – and they all followed. I wriggled through the eighteen inch pipe and then called for a volunteer to be my 'casualty' to take through the pipe with me. The biggest man volunteered – they obviously had it in for me – so I pulled a wry face and said, 'Just my luck to get him', whereupon the smallest man of them volunteered. Saying, 'That's better', I set to work and eventually after a great struggle I hauled this little fellow out of the far end of the pipe.

There were some ribald cheers and half hearted clapping – they were having a great time at my expense – whereupon I turned to the big fellow. 'Now, it's your turn', and I dragged him through the pipe so fast they were all astounded. Now there were cheers – real cheers – and for the remainder of the three days they ate out of my hand as good as gold – seniority all forgotten.

There followed a session with ropes and they learnt or re-

Carrying stretcher through 30" pipe.

Stretcher carry completed.

Casualty lowered on stretcher by three point suspension

learnt how to bend ropes for different purposes, always bearing in mind that human lives depended upon the accuracy of those knots. For my part I learned to tie right-handed, left-handed, towards oneself or away and also to do so blindfolded. I emphasised that it is only in the dark that the hands take over and dexterity is won. I reminded them that each of their men had to be able to do all these bends in the dark and be correct every time. I was able to stand behind them and move their hands for them or to stand in front and do it opposite handed so they could follow as if watching in a mirror. They began to feel a sense of achievement.

Dr Evans took over after lunch and gave them a session on first aid in wartime. There was much to learn.

Later I put them through all sorts of drills with stretchers and slings out of windows and through trap doors. I talked to them about the structure of wreckage and how blast played tricks, sometimes sucking out the walls leaving the floors to pancake making sandwiches of the furniture, and on others to throw the walls inward onto the furniture. At the end of the course everything was brought together in one or two incidents, with people wounded and suffering amidst debris, and one or two apparently missing. I taught them to search and to crawl in underneath the debris. Horror upon horror – a fire broke out. They rescued one man only to find half an hour later when someone was still missing that several of them had been standing right on top of the place where he lay buried!

I called them all together for a final chat and asked them whether it was essential for their men to continue their rescue training or to do other jobs? They were absolutely certain of one thing – that each of their deputies should be submitted to a similar exposure to the facts of life and death. They were agreed upon another point that the 'wounded' casualties were a must to give the maximum impact. Obligingly, I agreed to their instructions and fixed a whole week for the deputies a few days later (11-15 July, 1942). The candidates were younger, and there was more time and the course was tougher, but equally successful.

4

A Competition for London Region, Group 9

The Byfleet lady was home again for another special occasion, when our London counterparts requested us to stage and judge a rescue competition for heavy rescue parties from London Region, Group 9. This seemed to be both a challenge and a very special compliment, because London's status and experience were high.

We invited Byfleet Ambulance to help us with the casualties and the lady came too, and thus Brenda Whiteley made her debut at Leatherhead – a day to remember. It was decided to locate Brenda under a collapsed floor, beneath a beam that had broken her leg. With her came her dog Fifi, a husband and some other people. She was scared by her predicament – the dirt, the dust, plaster, soot, the pain in her leg, the terror of being held in a vice by the beam. Her dog had run off or was, perhaps, another victim. Her husband did not come, when she called. She had renewed pain when the beam was lifted off her leg, was profoundly grateful to her rescuers, frantic about dog and husband, and distressed that lunch would not be ready when the children came home from school. The rescue men had entered a new world. Here was a person with problems as well as injuries, and there was a need to care for the person as well as for the broken leg. This person was part of a community with responsibilities – not just another statistic.

Dr Evans, George Waters, G.M., my oposite number in Group 9, and I were the judges and we agreed that Brenda Whiteley had added a new dimension to rescue training. The tests went on for five days and then, on behalf of the other judges too, I was privileged to present to Brenda a bouquet of roses. It was no less than her due, but an unexpected delight to her.

But much else came out of that experience. Years later Brenda was to tell her side of the same story. 'I wriggled into position under a heavy beam at the Rescue School at Leatherhead. It was a nice fit for someone of my size, but I had no difficulty in getting into position. The rescue instructor checked over my briefing and left me, taking his electric torch with him. Under the wreckage I could see nothing, and all I could hear was the perpetual falling of little bits of plaster amongst the debris. I tried over my acting, remembering the essential words. Then I settled down to wait. It was lonely and somewhat eerie. Suppose they do not find me? Suppose the beam slips? I felt very much alone. I tried to wriggle, but apparently the beam had slipped, because it was tight down on my waist and, as I wriggled slightly, bits of debris jabbed into me. I struggled, but could not get out. In my panic I may have cried out, but nobody came. Would I be crushed before they found me? Then suddenly the instructor was there. He reached for my hand and said reassuringly, 'Take it easy. They won't be here for a few more minutes. Wriggle about a bit to make sure you are comfortable'. I wriggled nevertheless and, equally suddenly, there was plenty of space and I could move easily. I knew I could get out by myself if necessary. My panic had left me and my body returned to my normal size. I was glad that I hadn't told the instructor that I'd panicked. It was alright now. Afterwards I wondered whether in fact the instructor did know all the time, because he always managed to arrive just when I needed critical reassurance'.

I treasure this story, because I was that 'instructor' and I did know what it was like to be left and tried hard never to let it happen to others under my care. Brenda's performance opened my eyes to the immense possibilities of mounting much far-reaching training – proper reconnaissance; handling of the full human being and not just the wounded body; utilising the resources of the immediate community and its environment. We were at war!

5

Do it Yourself at Home

Although in many ways the outstanding work of the rescue school started with the course for Local Authority Engineers, important events preceded that traumatic occasion. Even while the two schools were running at Wonham Mill, Betchworth, and at Guildford, concern was being expressed at the great distances between the major urban areas where the official rescue parties were based and the distances within the rural districts and between the outlying villages. This was even more worrying, because those official rescue parties were only part-time and could take hours to mobilise.

The County Council, therefore, appointed seven full-time rescue parties based within the rural areas. These it was my privilege to train and through the rescue school to lead. Even this did not provide adequate cover for the villages. In an emergency the CD wardens, including the women, would do what they could. It was against policy to use women in rescue parties, but in emergency, if their men were not there, they would do what they could and they would be less likely to damage themselves if they were taught how best to do things. Towards the end of April, 1941, at least a month before the parachute mine fell on St Andrew's Convent at Leatherhead, the County CD Training Officer came to see me to enquire whether I would be willing to take on this essential warden training. They would have no equipment other than their own domestic resources and a few things, like ropes and perhaps a stretcher they could borrow. I accepted the challenge in the spirit in which it was offered. As Churchill promised, 'We will fight in the villages, in the fields and in the ditches'. I was already at war.

We had not yet learned about total destruction of hundreds of houses built of bricks and mortar and timber and tiles, so I taught them to use ropes and improvised tackle, how to secure an injured person to a door or stretcher, how to carry a casualty up or down steps and slide a stretcher down a ladder. The first two courses were held on the green at Shalford, two miles south of Guildford on the 3/4 May, and 17/18 May, 1941. During these courses on behalf of the Guildford Rural District Council friendships were formed which were to bear a rich harvest when the time came for Casualties Union to be formed. Each time I was entertained to lunch by Chief Warden Burgoyne who lavished some of his precious wines upon me and even scarcer rations. How generous people were in sharing whatever they had got.

At Leatherhead, as soon as we had patched up the chapel as a lecture and assembly room and a couple of rooms for the commandant and for the instructors, as well as a WC, we were sufficiently in action for the third course for Guildford Rural to be held on 22/23 June, 1941, with the first day being held at Shalford, while the second day was held at the school at Leatherhead.

This pattern continued until the end of 1944. I would go to the local authority for the first day and they would come to Leatherhead for the second. As many as two hundred and fifty trainees would come and in order that they might be subdivided into groups of less than ten, so that everyone got real practical experience, I would deploy three of my county rescue parties to act as instructors, each with a set task to teach. Every twenty or so minutes the trainees would change stations (like a whist drive) until every group had carried out every one of the exercises. I roved over the whole, chipping in here and there so that every group came under my influence, for better or worse, during the course of the day and to provide that essential continuity with the day before. It was an exhausting period of my life. All the week long the school trained the full-timers; in the evening I visited the part-timers at their home bases; at the weekends two days with the wardens, and once or twice a week I would visit the county rescue parties on standby duty at night. My identification with them and their problems, aided greatly to the thought process that went on in me continuously all the time, enabling the training patterns and the school facilities to develop in parallel. What wonderful people these my fellow citizens were; I was proud to be associated with them and even more to count them among my friends.

Lowering a casualty on a stretcher improvised from a chair and two poles.

Wardens lifting a 14 ton block of concrete with two levers.

Introductory chat with newcomers to the school.

Launching casualty on improvised stretcher through narrow window opening.

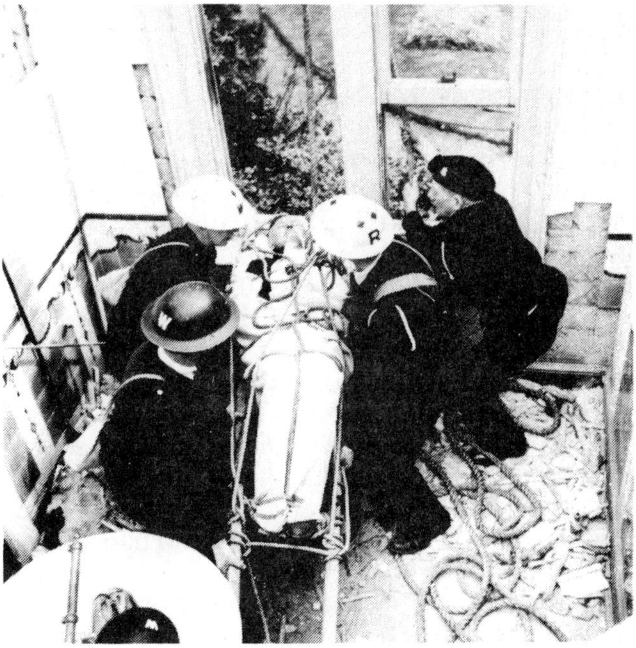

The men went on night duty at 7 pm until 7 am and they used to have their evening meal about 8 o'clock, so as a rule I was careful not to visit them until after they had had their meal break. One evening when I called at the Beare Green depot of Dorking and Horley Rural Division, I found that the leader had been dissatisfied with some element of training and had been giving his men some extra drilling to get it right, before allowing the food break, so they were about to sit down to their meal when I arrived. Please would I join them? Adams, a lively septuagenarian, was orderly that evening and said to me, 'Lucky you came tonight, sir, because it is so-and-so's night off, so you can have his ration'. I knew how the rationing worked and that there would be no ration for the man on his night off, but Adams was determined to make a plate for me. The meal consisted of two sausages and mashed potato with a full round of fried bread. When Adams set out the plates they looked most appetising.

We all sat down and something made me say to Adams, 'May I swap plates with you, please?' Reluctantly he acquiesced, and the dear generous old boy had covered each man's sausages with the fried bread, but on the plate he had served himself, there were no sausages. I could have cried at this magnanimous gesture, but we kept it a secret between the two of us until now, when I feel at liberty to pay my warmest respects to his memory – a lengthsman, in emergency turned rescueman – the salt of the earth. As a lengthsman you could never get onto his length of road on a frosty night before he had gritted it, nor could you find a pothole, nor a puddle, there for two days running, nor overhanging hedges, nor loss of visibility through cow parsley, or long grass. It was his length, and he kept it safe for everybody. Such fine men may well be equalled, but they can never be surpassed – a very fine breed. God bless every one of them.

Following on the heels of the pioneers from Guildford Rural, the wardens from Caterham and Warlingham Urban were eager to come to Leatherhead, and on 8 February, 1942 and again on the 22nd they did and then six weeks later a third group came on 11 April. Each time they had had a preparatory talk the previous day, when I visited them. This was beneficial, because it enabled me to gauge the sense of organisation, or lack of it, at the home base. On the third occasion, with my

Casualty lowered on stretcher by sliding down ladder.

Casualty lowered on stretcher by bearers on two ladders.

R. Lovesy wearing protective arm and leg guards in broken glass.

consent, they brought a crack Press photographer from the weekly picture magazine, *Illustrated*, his name Jarché. He watched everything and photographed anything that took his fancy, without fear or favour. A week later a reporter from *Illustrated* invited herself to the school - Miss Marshall. With Jarché's excellent photographs and the penetrating questions of Miss Marshall, *Illustrated* published on 31 May a three page feature under the general title Battle School For Wardens. A proud occasion for all concerned. In the final analysis two thousand six hundred wardens took part in this course with a significant number from all over England who wrote requesting training after having seen the feature in *Illustrated*. These warden courses captured the imagination and I believe that it was they, more than anything else, that put me in line for the M.B.E. six months later.

It was probably as a result of the article in *Illustrated* that we received a request by Gaumont British, and subsequently by Pathé Gazette, for their newsreel cameras to visit the rescue school to record some of our students at work. This was very exciting and we ensured that an interesting programme was in progress while they were there. Our best casualties, as well as some competent rescue personnel, were present, and the camera teams were enthusiastic about the material they had recorded. They were shooting for upwards of an hour, in each case, so we were encouraged to expect a short, but informative, news item on our activities. However, our hopes were dashed. The editor decided, that it would not do to display such 'horrific' material on the public screen as it might cause pregnant women to have miscarriages in the cinema! It seemed a bit far-fetched to me but, who knows, the censor may have had something to do with it as they were often cagey about all aspects of the war effort. So we got only some thirty seconds of the most innocuous bits of their films. Both companies showed me their films, and the excerpts they were permitted to show. Disappointing. But, just one of the fortunes of war!

On 16 September, 1942, we were granted the cockpit of a fighter aircraft so we could teach all our students how to rescue a fighter pilot from a crashed aircraft. We were proud to be able to give such positive help to those who were keeping the enemy at bay, and one can only hope that our efforts were indeed fruitful. There was no way of knowing that the skills we gave our students did actually save lives, but we found our own hands becoming more sure and more deft.

6

Getting down to the real task

The first course for full-timers at Leatherhead commenced on 21 August, 1941, and continued until September, 1942, embracing some forty rescue parties. The course was different in several ways from the type of training that was possible at the previous school sites. The important things that became possible were practice in reconnaissance and searching for casualties; handling casualties in confined spaces. All courses incorporated these elements. However, once the first course for Heads of Rescue Services had been held, the use of made-up responsive acting 'casualties' pin-pointed and, indeed, focussed attention on the casualties themselves. The danger and difficulties of the situations amongst wreckage was taken in their stride, for the rescuemen were intent on handling the injured person within these confined spaces, having regard to the injuries. Methods varied considerably depending upon whether there was injury to legs, chest, head, or two parts or, perhaps, more.

Dr Evans found it difficult to be available just whenever he was needed at the rescue school, but he sent his assistant, Dr James L Barford, in his place and we saw less of Dr Evans, but a great deal more of Dr Barford and the course expanded on the medical side to balance the tremendous progress that had been made on the engineering side.

Dr Barford gave lectures on the principles of first aid in wartime; shock; bleeding; head injuries and fractures. Meanwhile an intensive study was made of bandaging for all circumstances, recognising that bandages to stop bleeding had to remain effective during handling. In the upshot the students were called upon in one exercise, or another, to put dressings and bandages on every square inch of the human body, within the pattern of the course.

Dr Barford turned out to be a genius at putting over this type of training. With each of his lectures he used action as well as words. Almost all his time was spent examining and diagnosing the nature of what he found and applying treatment, or indicating what treatment could be applied and the effect it would have upon the injured person. Throughout his talks casualties would wander in at precise moments to teach the next point, or with major injuries, casualties would be helped in in an appropriate way. Brenda Whiteley was most frequently his casualty, though others, like Bill Warn, shared in this duty. The speed at which they recreated different situations, when serving alone, was truly remarkable, but it was easier and more effective if there were several casualties, so as to avoid any confusion in the minds of the students between one condition and the next. The students were entranced and although they never learned the Greek words so common in medical parlance, like haemorrhage, they became competent at stopping bleeding. It was simple, straightforward and practical having regard to the environment in which they were likely to meet such injuries, and the students were given the maximum possible practice in these measures under realistic conditions.

After three months to think it over and no doubt chewing it over in twos and threes, the Heads of Rescue Services came back at their own request for a further course in October. By that time the use of skilled, made-up responsive, acting casualties was becoming standard in all our training, but for this course the mental agony of the presentation of suffering and trauma was calculated to have a maximal impact. One incident that remains crystal clear in my mind, as I write, was a family caught in a basement shelter. Mother (Brenda Whiteley) was badly hurt with internal injuries, her baby was dead, and her husband could not be found anywhere. These poor inexperienced men got the mother out with some difficulty, having to stand for a long time at the foot of the steps, where a certain amount of debris had collected, as they manoeuvered to get the woman up the steps to safety. What were they to do with the dead baby? – mother was badly shocked and in great pain. She kept calling for her baby, trying all the while to get off the stretcher to go and search for it, and

Collapsing roof, during visit of forty M.O.H., casualty beneath. (SEE PAGE 71)

she wanted her husband – he would soon set things to rights. I remember one of the men saying in desperation, 'Oh! Let's give her the baby; she is too ill to realise it's dead'. And then – imagine their distress to discover that while they were at the bottom, struggling to get the mother's stretcher up the steps, they had been standing on the father who had been knocked unconscious and been buried at the foot of those steps. They emerged finally with the unconscious man (Cyril Wallis) and I was aware that his rescuers were in a cold sweat. They learned very fast in that hour, but I felt an absolute brute to have subjected them to that ordeal; but it was a lesson that they had to learn. Training of the Rescue Service was never again interrupted and perhaps I dare to add that they all became my firm friends and could not do enough to help the school. Need I add that the chiefs were eager that their deputies should tread the same road and the following week another course was held for the deputies.

7

Contest for Warden Recue Teams and Housewife's Aid Points

While this was going on during weektime, the training of part-timers and wardens was continuing in the evenings and at weekends. Bed was a luxury only to be found listed in furniture catalogues. The success of warden training and of the competitions staged for rescue parties had led to a demand for a suitable contest to be staged for rescue teams of wardens in association with housewife's aid points. On 3, 10 and 17 October, 1942, heats were run for a wardens rescue competition, and on the 25 October the final was staged in the presence of the Rt Hon Viscount Monsell, G.B.E., Regional Commissioner Region XII, Sir Philip Henriques, K.B.E., Chairman of the Emergency Committee of the Surrey County Council and Capt. L Murray Robertson, Surrey Civil Defence Controller.

The arrangements, which included creating stands of scaffolding and planks for the hundreds of spectators, were in the hands of Capt. R Lovesy, the Officer-in-Charge of the school's administration. He was helped greatly by the County Divisional Road Engineers and the county rescue parties. The incident was of a remote village cottage bombed with fire and high explosives, where the wardens could not expect rescue parties to arrive under thirty minutes. The incident started with the explosion of a phosphorus bomb (commonly called a molotov cocktail) only a few feet away from where our principal guests were seated. The bomb spattered fire very close to but not actually hitting a serious casualty spread-eagled on the ground. Three teams in turn were called upon to handle this situation, which was repeated precisely for the second and third team.

The situation was created and developed according to plan for the first two teams and duly dealt with. When it came to the third team the spatter of phosphorus proved to be entirely different and the casualty's hair caught fire. Instantly our 'casualty' protection system took over, the contents of a bucket of water hit the casualty's head extinguishing the flames and the casualty was whipped off to an emergency station (my office), where Dr Barford did all that was necessary with a solution of copper sulphate, and the casualty, although surprised, was unhurt.

The explanation for this mishap was that the operatives were so cautious in rehearsing the scatter of fire from a lemonade type of bottle containing phosphorus liquid, that they failed to leave three untouched bottles. They had, therefore, used one with a gingerbeer type of bottle, which had thicker glass and with which they had not practised. When the third bottle was exploded, like the others by a slug from an airgun, the thicker glass threw the contents just that much further. It was the judges, of whom the writer was one, who had the most difficult task, but clearly the team in question acquitted themselves well. The spectators thought it was a great show and clearly there was much first class training being undertaken by all the participants.

8

We go into print

It had never occurred to us that other people were not doing exactly the same sort of training as we were, but after the publicity given by the feature in *Illustrated*, requests began to arrive for more information about our principles and methods. We wrote notes covering the whole of our training for our visiting students to take away, and these we distributed as requested, but the real demand was for the training of all wardens to do rescue work with improvised equipment. So in the autumn of 1942 I was at last persuaded to write a simple and well illustrated book, *Practical Rescue Training for Civil Defence Wardens* and Dr Evans generously contributed a timely chapter on *First Aid in Wartime*. By the time the bulk of the book had been written the acting-made-up casualty had appeared on the scene so I was able to include a chapter on 'Live Ammunition'. Pitman published it in spring 1943.

Again this chapter whetted people's appetites and requests came for more information. We gave them the instructional leaflets, but these were not sufficient so Dr Evans and I got down to it once more and wrote a second book, *Battle Training for Civil Defence*. Lanta Spurrier, herself a warden, designed the cover and it was published by Gale and Poulden in spring 1944, just one year after the first book.

The demand for skilled casualties was becoming tremendous. I had become one of them because I valued the experience of being handled by the students as a means of assessing their true ability. Clearly the dozen of us were being greatly stretched. The others were all volunteers, living a distance away from the school, and during many weeks we were asking for their services all the seven days. We needed to enlarge the group. The problem was how?

There was however a new aspect that came to light, of which I myself was fully aware before the others started to tell me. In

Walking casualty helped down after dressing and rescue.

studying the situation to be enacted, the signs and symptoms to be envinced, practising the control of one's body to enable one to show the loss of power, mobility or function, eg with broken or dislocated bones or joints, one learnt a great deal more about reconnaissance, diagnosis and the type of treatment that helped, comforted or aggravated one's problems, mental as well as physical. In short the 'casualty' learnt MOST. The experiences were challenging, satisfying, revealing and above all memorable.

We wanted more help as 'casualties', but if the 'casualties' learnt most clearly it was our duty to offer this means of training to the whole of Civil Defence. If in the process we gained ten to twenty recruits, it would reduce the immense burden that the small nucleus was carrying at that time.

I drafted a letter to all local authority CD Sub-controllers, which I eventually persuaded the County CD Training Officer to send out on the 9 November, 1942, calling a meeting at the rescue school, Leatherhead, on Sunday, 22 November from 10 am – 5 pm. He was sceptical, believing that good as our training was, we must all be barmy to allow ourselves voluntarily to be subjected to such dirty and dangerous experiences.

9

Casualties Union is formed

The 22 November, 1942, was a bright and clear morning with a heavy frost and mist on top of the Surrey hills. Our team of demonstrators arrived from Byfleet and Woking; the duty rescue party lit fires in appropriate places especially in the canteen and the long hut. These two places were expected to be more than adequate to accommodate our visitors and staff. We were happy that the accommodation was warm when the first of our guests arrived. I was glad to see among them many faces I had seen during the warden rescue courses, with whom I had already created a valuable rapport.

We started on time. I welcomed them, told them something of our programme and introduced Brenda Whiteley to them. Brenda had to stand on a chair to be seen over their heads. I held onto the chair because Brenda was inspired, telling them of the value, the interest, the importance of this work both to the individual and to the community. The crowds kept arriving and pushing into the room. Brenda introduced Dr Evans, who added medical authority to the views that Brenda had already expressed. In turn Dr Evans introduced Dr Barford, who told them something of the value he found in the use of skilled 'casualties' to illustrate his lectures and the extra power it gave to presentation understanding and memorability.

I invited the party to come outside to witness a demonstration incident involving some half dozen casualties. We were met by an equal number of visitors who had arrived late because of the mist on the hills. I bid them welcome and directed their attention to a distressing situation in another part of the complex, where a man in his bed found himself precariously suspended high above a flight of steps leading down to the basement. Rescuemen were already climbing up the remains of the building to reach him. Others were searching for other members of the household who it was feared might be casualties. The situation was both dramatic and inspiring. While some of the men managed to haul the stricken man on the bed (A George Evershed) to safety, other casualties were also being led to safety (Edward Akester, A L Andrews, C O Hankin, Mrs O Elspeth Bristow and Mrs Brenda Whiteley). Some were hobbling along with help. Others with face and head injuries stumbled after them. This enabled the audience to see the nature of their injuries and their limitation of movement – a soul searching sight and a bit sick-making for some. How they would be able to cope with such casualties on their own patch was a thought in the minds of many.

This was followed by a demonstration of how the apparent injuries were created. Akester, the artist, a cabinet maker by trade, made a remarkable recovery to undertake this task. Bitterly cold though it was, Brenda and others submitted to being 'wounded'. Brenda showed how the pallor was applied to the face to indicate shock commensurate with the injuries and where appropriate blueing the lips and creating that sunken look around the eyes. She showed the way it was necessary to work the colour up into the hair, behind the ears and over the throat and neck.

I then showed them something of how the staging was done so that it remained comparatively safe and also how, if there were any risks involved, a special watch was kept by a 'prisoner's friend', who would never be out of sight or reach to take immediate action as necessary. I reminded them that by this technique, if accurately portrayed, it became possible to observe the signs and elicit the symptoms of the victim's condition so that an adequate diagnosis could be made for the victim's protection during movement, and also for the provision of immediate medical aid.

The reception was so warm that I began to visualise the difficulties of providing the necessary training for such a large gathering. I was already in difficulties with catering for three hundred and fifty visitors plus my own crew – some four hundred plus against my estimate of one hundred. The WVS were marvellous and the food controller treated this as an emergency feeding exercise and helped as much as he could. There was hot soup, bread and tea to warm and cheer us up and we were also helped by a number of wiser guests, who had brought their own rations.

After lunch I explained the sort of training we had in mind and stressed that we expected everyone to do everything without question, including being trapped in debris. I emphasised all the risks – perhaps over emphasised would be more correct – the dust, the grit, the jagged bits of broken timber, the inevitable broken glass amongst it all and of course the ever present risk of the collapse of the wreckage due to somebody climbing over or pulling away the wrong piece of debris. There were also the bumps and bruises that were an inevitable part of being a 'casualty'. Members of CD were officially insured on training duties and service as a casualty would be covered by this insurance.

At the end of the day, after the type of organisation we had in mind had been further explained, visitors were invited to enrol in 'Casualties Union'. Amid considerable excitement, one hundred and twenty volunteers enrolled. It was a wonderful result and, quietly, a relief to me that no more would need immediate training. They were, however, to increase to some five hundred before the end of the war.

Starting a fortnight later, two training days were held on Sundays, 6 and 13 December, 1942. And a further two Sundays, 3 and 17 January, 1943. On Sunday 14 February, the new members were invited to experience being rescued from all sorts of places by all sorts of method so they could get accustomed to many manoeuvres they had to expect. The basic training was simple. Each session started with make-up of wounds and types of injury to be studied. First day gave emphasis to the acting of fractures – pain, loss of movement and power, deformity and possible positions of least pain, automatic self-protection and self-splinting. This was followed by consideration of appropriate incident planning with the realistic touch. All untoward situations, whether small or great, were called 'incidents'. Second day studied traumatic amputations, their make-up and presentation. Later we rehearsed our reaction to handling of injuries in general with response to reassurance or increased terror under adverse treatment. Third day studied burns, skull injuries and asphyxia. The staging of incidents, wherein such injuries could have occurred, would need to cover protection from residual fire. Fourth day brought a study of internal and trunk injuries. We also considered the staging of incidents to exercise wardens, first aiders, rescuemen, firemen and police. It was a sketchy enough beginning, but the teachers themselves still had a lot to learn, because it was all so new. Brenda Whiteley, Edward Akester, Dr James Barford and I carried the brunt of this pioneer work.

All these sessions were run on the whist drive principle, where after a standard time at one exercise, the participants moved on to the next. In this way with limited objectives at each station, it was possible to keep the training to small groups receiving individual attention, so that no-one, however shy, was left out and everyone had a go at everything.

And so the rescue school dropped into a new gear, and was able to carry the ever increasing load.

10

Battle training begins

Inspired by the article in *Illustrated* in May, 1942, we were now ready to launch a course entitled, Battlefit Course, for full-time rescue parties. In this course first aid was to play a full part and Doc, as we called Dr James Barford, undertook a series of lectures and training exercises associated with each subject in turn. He would talk about first aid in wartime to emphasise that under battle conditions objectives had to be simple and direct – stop bleeding, maintain breathing or restore it, prevent further injury or aggravation of those already endured, immediate postural treatment and, where appropriate, priority for removal to second aid.

His talk on bleeding was illustrated by a series of casualties; one would come in bleeding profusely from a gash on the wrist and as Doc raised the limb the bleeding became less and less. With the same casualty he was able to show several places where digital pressure on the artery stopped the bleeding; how a pad and bandage firmly applied would check the loss of blood, and then as the bleeding seeped through the dressing he would show how a second dressing applied over the first could succeed; or, perhaps, the combined effect with the raising of the limb would control the bleeding. Another casualty would have an ugly and large wound far too big to be covered by one dressing and he would show the difficulty in staunching such a wound across the shoulders and chest, because although several field dressings could be made to cover the area and be firmly bound into position, as soon as the casualty's position was changed the bandages would become either loose or suffocatingly tight. Either way they would need to be adjusted. Yet another casualty would have internal haemorrhage in the chest; another in the abdomen and a third in the head.

Of the circulation generally, he explained the function with a cunning little demonstration consisting of a pump drawing dirty looking 'blood' from one glass container, passing it through the oxygenating process in the 'lungs' and delivering it bright red into a second glass. The simple secret – a dilute acid mixture in the second glass and an alkali in the first with a litmus solution in the mixtures.

Talking of shock, Doc liked to illustrate the effects by a casualty suffering from delayed shock. Uninjured and apparently unscathed, the casualty would appear wandering vaguely, apathetically, shrunk in stature, unable to concentrate, blurred of vision as well as of mind, barely breathing, completely exhausted and in imminent danger of collapse. This would be mental shock, but he explained that every casualty, however minor the injuries might be, would suffer some element of the severe shock they had just witnessed, and the importance of gentleness and positive reassurance by actions as well as by words, and preventing blood loss.

The students found Doc's lectures lucid and unforgettable and everyone of them understood the ways to save life and above all to stop bleeding. By the time they had passed through a week of training under Doc's guidance, they would have to place a dressing in every part of the body in one exercise, or incident, or another. All the difficult places were included.

The students arrived (thanks to records of sirens, bombing and other air-raid sounds, helpfully supplied by the BBC) as bombs fell in the vicinity of the rescue school and before they realised that they had come to the right place, they were dealing with an emergency. Without previous training, they met many problems for the first time the hard way, but half an hour later they were relieved to discover that this was only a first lesson at the rescue school, and their training proper was about to start.

I had the privilege of welcoming them to a very tough type of training and then explained to them that the strategy for dealing with such ugly situations could be built round the four letter word L-I-F-E. This has been referred to. They worked with ropes. They had anti-claustrophobia training and later we attacked the fear of heights, until they had walked along a four and a half inch wall thirty feet above the ground with a sheer drop. If I was conducting this exercise, I allowed the student to hold the tip of my little finger, while I walked along a parallel four and a half inch wall some four feet away. Amongst the hundreds who did this exercise only one slipped and I was fortunate in being able to grab him by the wrist and haul him back to safety. After that we walked the wall again.

Everything was positive and had a positive purpose in the pattern of skills we were seeking. The students in their brief time with us needed to learn to make an adequate diagnosis and to apply positive treatment. They needed to learn how to ask the right questions and to take full notice of the replies – the casualty might not be able to repeat. The students had to learn also the priorities that followed from these diagnoses – the priority for attention, the priority for removal and at times the priority of obstruction when other casualties, who might well be in more urgent need, could not be reached until the less severely injured had been removed. The course provided opportunities to face all these priorities and make decisions. In one oft repeated incident Brenda Whiteley bleeding profusely was located temptingly high up on a pile of wreckage. The students immediately set about dealing with her, only to discover later that another casualty, whose existence they had been led by the warden to expect to find, but whom they had been over hasty in not seeking, lay buried unconscious in the wreckage beneath their feet. Imagine their horror in finding him and that that much trampled on casualty was none other than the commandant – myself. In such ways their experiences were made positive, cumulative and unforgettable.

Every incident staged throughout the course was designed to highlight one or more essential elements in dealing with wounded people in an air-raid situation. It was essential that in association with the local wardens, or in their absence, to make enquiries as to how many people were likely to be involved, and anything about their habits that might help to locate them. While the leader was doing that, the rest of the party surrounded the area and listened for some response to their calling 'rescue party searching', and looking for ways in under the wreckage, whilst taking care to avoid clambering on top. Once a casualty had been located, careful questioning and attention to the responses could lead to the finding of others. So the search and rescue proceeded. The students were being stretched and trained at the same time by positive

Casualty marooned on upper floor.

experience. Without the help of skilled and highly disciplined 'casualties' such training would not have been possible. War gas (CAP), real incendiaries and phosphorus bombs were woven into the pattern so the experience was extensive.

To end the course they were given a demonstration incident undertaken by Surrey County's own skilled rescue parties in combination with warden, ambulance and police services from Leatherhead. It becomes obvious that the success of the rescue school owed much to the faithful support of its neighbours and friends.

11

Testing the Pongophone

Towards the end of December, 1942, Regional Control requested the school in its research capacity to set up an effective test for a highly sensitive device for seeking casualties buried amongst wreckage and debris. It was considered that to appraise the equipment the tests should be graded from comparatively easy to exceedingly difficult in the hopes that the limits of usefulness could be defined. No information was given as to the nature of the equipment so that the school had to create tests for a wide range of possibilities.

The five hundred ton pile of debris with all its cavities was used. The heap had been constructed upon a spread of concrete formerly used for netball in the days of the convent. It was thought that this might be a useful conductor of sound; the slab was therefore broken up around the whole perimeter of the pile and the broken areas were covered with extra debris. The plan was threefold: one casualty would be free within one of the cavities, a second was to be smothered with two feet deep of sawdust within a cavity; the third was to be enclosed in a corrugated iron 'coffin' cushioned on rubber pads and packed around with sawdust. Each test was to comprise three stages: groaning; calling for 'help'; and tapping. The 'coffin' was inserted within the adit created by a tunnelling exercise that fortuitously had just been completed at the time of the request and had not at that time been restored. Eight sacks of sawdust were dragged through the anti-claustrophobia crawl to the cavity chosen for the tests. A special entry to the tunnel was created from an adjoining building, so that the 'casualties' could enter and leave unseen. A speaking tube was finally threaded through the cavities so that the administrator could keep in touch with the 'casualties' without being overheard by the operators of the detector.

When the time came for the tests to begin, the three of us

'casualties', Brenda (forty-five), Doc (say sixty) and myself (thirty-three), crawled into the complex through the hidden entrance and Brenda slipped into her box facing the cavity so she could watch. Doc and I squatted on the floor and awaited a call from a directional nautical loud-hailer. We could hear it very clearly. Dutifully I groaned every few seconds; after about five minutes I began calling, 'Help' – at first as a wounded person but finally as a lusty rescueman trained to give commands on the parade ground. After some five minutes more I began to tap – on the ground; on wooden and then on metal debris for another five minutes.

Immediately Doc lay flat on the ground face down. I put a clean towel over his head and he made a cavity around his nose and mouth with his hand; then I covered him all over with a sheet, head included. This was difficult enough because of the closely confined space but next I had to tip four or five sacks of sawdust over him so at to fill the cavity up to the roof – some two feet or more of sawdust. One of his hands was left exposed for silent contact with me. By messenger from the other end of the speaking tube we were able to alert the searchers that test two was ready. Their sole communication with the 'casualties' was by the directional loud-hailer as used at sea between vessels. It was very powerful. They called for Doc to groan. We were certain that he could hear because both Brenda and I had practised under the same conditions. It must have been very quiet groaning because even we could not hear; eventually he groaned so that it was audible to us. When it came to the time to change, the loud-hailer asked for the casualty to tap. Doc continued to groan but eventually he tapped. Then came the request for him to call for help. This was bound to be difficult because of the restricted air space, the fine dust and the heat and weight of the pile of sawdust on top of him. I was becoming concerned for his safety because although I had been so buried for forty-five minutes on a previous occasion, I was some thirty years younger than Doc. I uncovered him rapidly – that is as rapidly as I could in that confined space – and he immediately sat bolt upright and, pointing an accusing finger at me, said, 'You panicked'!

Then it was Brenda's turn. She had been lying still inside her box, able to watch the antics of her two companions. Now a lashing was attached to her wrist passed through a hole in a wooden end that was to complete the box and the other was tied to my wrist. The end was placed in position, signalling checked and the end backed by two feet of sawdust. Doc and I could still see each other around the pile of sawdust, though neither of us had much room to move. The process of groaning, calling and tapping was repeated without mishap and Brenda was released. We prepared the facilities for similar tests to be continued on the following day and a cramped but triumphant trio of 'casualties' or 'guinea pigs' emerged to be warmly thanked by the Regional Commissioner, Lord Monsell, and the CD Controllers of the local authorities within the Region. The Regional Commissioner took the opportunity of congratulating me on the award of the M.B.E. in the New Years Honours List. Doc presented me with a sailor's cap with, 'Mud, Blood and Enthusiasm', on the ribbon.

The tests continued in an orderly manner for a further two days, but the machine proved unhappily to be insufficiently sensitive, try as diligently as they did. The engineers of the local authorities attended on the second day and the CD Officers on the third day. A great challenge for the new Casualties Union, when one realises that the recently enrolled membership had not then completed their initial training. What a grand start!

12

Truly a battle school

The reader will appreciate that those who were drafted to the rescue school faced a pretty gruelling week of learning and endeavour and above all self-control. It takes a strong man to be gentle with injured people under such conditions. As the weeks flew by the fighting service asked to be included in our programme and in April, 1943 the first soldiers joined in the course being run for the CD Services. The numbers grew week by week until in February, 1944, there were thirty to forty servicemen each week until hostilities in the Far East ended in June, 1945. During this period there were all sorts of special courses for military personnel. In August, 1943, a special fourteen day course was held for eight Officers and thirty-three NCOs prior to exercise Harlequin, which could have been the start of the invasion of Europe. Before that, the exercise, Luck, made Leatherhead the scene of an invasion exercise, in which Home Guard, military, police and Civil Defence were involved, while Casualties Union provided wounded people throughout the combat area.

Providing so many badly wounded casualties was very heavy on our personal clothing, because each wound needed rips in appropriate places and of course amputations required even more preparation starting with the clothing. From the beginning of Casualties Union, we begged old clothing and Brenda Whiteley became our wardrobe mistress, carefully husbanding the various items so they could be allocated to particular injuries and according to whether the casualty was male or female, small or large. From time to time these precious pieces of apparel needed washing. In spite of the fact that blood coloured dyes did not always come out too well, our diligent Brenda kept the wardrobe hygienic. Inevitably there was a need for soap – a commodity in short supply and already rationed. The generosity that met our appeal was heartwarming for we needed the soap not only for our clothing but for our own bodies after serving as 'casualties'. I recall one touching assignment from a County Divisional Surveyor and addressed, 'For Lovesy's lovely ladies'. It was perfumed toilet soap!

Readers will have gained the impression that our activities, especially in the preparation rooms, would create a 'bloody' mess. This could indeed be so, but our members' tradition started at the outset to leave the site of operations and the preparation rooms perfectly clean and tidy and totally without litter. It was a healthy example assiduously followed by our loyal and devoted membership.

Another touching tribute came from a Caterham and Warlingham warden named Chinnock. It depicted by statuette, a nonchalant demonstrator balancing on one leg on a precarious pile of bricks on the topmost part of a wrecked building overlooking a fearsomely spiked iron railing and a terrified warden looking up at the demonstrator saying,

'Oh! Mr Claxton, I have a wife and seven children. Must I do that?' 'Yes, of course. There's nothing to it. Why I used to be much worse than you - come right up beside me. By the way, if you do fall, I should try to avoid those spikes. They were sharpened yesterday. We put them there to make you fellows careful'.

The little conversation was engraved at the foot of the wall in the little statuette.

I recall many occasions, when my original instructors, Bullen and Thomas, who were some fifty years of age, saved up the heavy or awkward students, until I was available and upon whom I was said to be expected to demonstrate fireman's lift, because every group said, sooner or later, 'How do you cope with a chap like that?'

'The same as with anyone else', my instructor would reply. 'The gov'nor will be along shortly and he will show you'.

I admired their confidence in me - a confidence I did not wholly share. I weighed 12st 6lbs. My first real test was the leader of the part-time rescue party at Ockley, on the A29 south of Dorking. We were meeting in the cricket pavilion on the village green, when the challenge came. He was stocky and a little under seventeen stone. He did not bend very easily, but I got him up onto my shoulder all right and straightened up;

my back was not very keen on it and my legs felt like jelly, and the floor of the pavilion bent ominously under our combined weight, I was glad that I was not high up and about to take my friend with me down a long ladder. Of course such entertainments were bound to come eventually – fortunately later and at the rescue school, where such challenges did arise.

There was a barrow boy from Purley who seemed to be just about as broad as he was long, with a couple of feet somewhere down below. He had no waist – just the equator with a leather girdle. He said he tipped the scale at between twenty and twenty-one stones. When I got him on my shoulder, I reckoned that he would crush the normal scales, as he nearly did me. A nice person to lift and to carry by fireman's lift, obligingly drapes himself, albeit unconsciously, and totally limp across one's shoulders, where he balances perfectly, but my bonny barrow boy rolled about like a giant marble and it took all my skill to keep him secure with the aid of his arm. Once more I was glad that neither window nor ladder were involved.

Time came when a party on an upper floor was doing ladder training, and someone said, 'How do you deal with him?' When I arrived, I obliged as custom had it I would. We had trouble with the small aperture of the window, because our clothing tended to catch in the tight space, but we managed to get onto the ladder. My legs did not like it very much, but I was happy that the rungs of the ladder were equal to the task suddenly put upon them and also that I had had plenty of practice upon people of normal dimensions.

Looking back over the years, I wonder which of these special loads caused me to suffer a dislocated vertebra in the small of my back in the years following the war. However, I am eternally grateful to the chiropractor, who sorted it out for me in 1951, so that twelve months later I was no worse than 'shop soiled'.

13

Special guests

The soldiers, normally peace-loving citizens, found the training at the rescue school, with its emphasis on saving life and limb, a very welcome change from their military training concentrating on the taking of life. Not surprisingly Commanding Officers found that men would volunteer more than once to attend courses at the rescue school. The fact that the school was a civilian establishment caused several COs to wonder whether their men were skiving at a cushy civilian school, losing the toughness they had been at pains to instil into them? One Colonel arrived without warning one Tuesday morning to my intense annoyance because the strict rationing to which a civilian establishment was subjected made even one unexpected guest a burden upon the rest. When chided he explained that he was anxious to catch us unaware lest we staged something special for his benefit. I told him he might go anywhere so that he might witness everything that was going on. At the end of the day, prowling from exercise to lecture to incident, he told me that he was most grateful for his visit and that he considered that the rescue school was even tougher than the military battle schools. He was proud that his men should be given such training. Approximately two thousand military personnel from many countries participated in our training.

On 19 November, 1942, we had a special privilege, a visit from the King's aunt, HRH Princess Marie Louise, who lived in our territory. Sir Phillip Henriques came also to welcome HRH to one of the County Council's establishments and then gave her into my care. We had staged an incident that was reasonably visible from a safe distance, but the grand old Lady was so interested that she insisted on getting amongst the debris to visit each casualty upon location. I was scared for her safety, but she was adamant, so I took her by the elbow, as I might have taken my mother, and together we invaded the wreckage to interview each casualty in turn. HRH was horrified at the situations, yet delighted that such training was being undertaken and promised that she would have great pleasure in telling the King all about her visit. After the incident was completed HRH requested that everybody working at the school should be presented to her and she congratulated each one and applauded their performances. It was a joy to receive such a charming guest.

So much happened so quickly; in April the CD Staff College at Stoke d'Abernon made a routine of bringing their courses to Leatherhead to learn about battle training for Civil Defence. Groups of MOH, CD Controllers, Allied Nationals came, about forty each time. Then came the National Fire Service. As a result I made several visits to the Fire Service Staff College at Roedean, Brighton, to talk to their courses and a number of senior Fire Officers came on courses at Leatherhead.

I well remember the day of the visit of forty Medical Officers of Health, drawn from local authorities all over UK and Northern Ireland. We had prepared carefully during the morning and I had directed exactly where they were to de-bus and I met them as they alighted from the coach on the less damaged side of the buildings. I led them diagonally across the site so they could see an orderly row of wooden huts when suddenly, as they came in view of the damaged area, there was a frantic shout from aloft. 'Look out below!', and an enormous smash as a roof crashed down. The air was full of dust; tiles flew in all directions; out of the dust appeared a woman, who was obviously hurt, shocked and breathing with difficulty. She staggered towards us – it was Brenda – full of apologies that somehow she had upset the whole afternoon's programme. Her head appeared to be bleeding. She said she was very sorry! I tried to show a mixture of annoyance and sympathetic understanding, telling her to go and get herself patched up and have a rest until she felt better. Suddenly I remembered that all my guests were doctors; what better than that a couple of them should attend to the injured woman. Two of the visitors immediately detached themselves from the party and took charge of Brenda. I piloted the remainder to the lecture room where I had proposed to give them an address of welcome and paint the picture of rescue work and the necessary training, before turning to practical demonstrations.

The guests commiserated with me that such an untoward occurrence should have happened during their visit to the school. I accepted their sympathy and said how grateful I was that their colleagues had taken one nasty problem off my plate. It was ten minutes or more before the two volunteers rejoined the party and explained to their companions that it had indeed been a training exercise to involve them; not an accident at all. It had, they said, taken them some minutes before they realised that the woman was acting the part of a distraught and injured person, who was going to get one hell of a roasting after they had gone. They were very impressed and so was their audience. So I took the rostrum and gave my talk.

We showed them a number of exercises concerned with rescue – search, emergency first aid, extrication and disposal. We had a break for tea and a biscuit. While the guests were otherwise occupied, my chaps were preparing a second disaster. We were intent on demonstrating tunnelling through bomb debris, when all of a sudden there was a strong smell of town gas. In the days before North Sea gas, town gas had a strong and characteristic smell and everyone feared an explosion, or perhaps, worse, asphyxiation of the chap who was working at the tunnel face. By the time that everybody had gathered round, we saw a man disappear into the tunnel; he was trailing a rope and a long length of hosepipe behind him; one of his mates was guarding the other end of both rope and pipe, ensuring that the end of the pipe was always in the fresh air. The rescuer had disappeared so quickly that there had hardly been time to see that he was wearing a facemask; the pipe and the rope were fixed to his waist; the mask was connected to the pipe by a flexible tube like that used in the war service gasmask. Our guests were very impressed by the smell; our prohibition of smoking; the rescue and the limp casualty. It was Bill Warn, whose face looked exactly like a case of carbon-monoxide poisoning. They approved the artificial respiration, but the doctors had no feeling for the victim – he presented no appeal to them. They remained unmoved. It was a good demonstration, they said. I wondered whether we had called their bluff once too often? At the right moment I enquired how they had known without a second glance that the victim was not a genuine case of carbon-monoxide poisoning? They affirmed that his behaviour was good, the smell of gas convincing, his face was absolutely right, but his wrists and fingernails were perfectly healthy. We had been found wanting; our presentation was not good enough. I thanked them for this timely lesson.

I had been briefed by Capt. Shakeshaft, Commandant of the Staff College, to instruct these doctors in the way to teach first-aid. It had seemed to us rather like teaching one's grandmother how to suck eggs, so we planned to interlace demonstrations of our methods of teaching with two incidents, which we hoped would appear to be accidents - well separated in time and utterly different in conception. Our guests had obviously learnt a lot from us, but we had learnt a very salutary lesson from them. It had been a very good day for the rescue school.

14

Light rescue parties

In January, 1943, we were requested to train first aid parties who had been reorganised as light rescue parties. They had very little rescue equipment. Their revised brief was to gather readily accessible casualties from the wreckage and care for their wounds rather than just attend to wounded brought out by the heavy rescue parties. They needed to be trained in handling injured persons in confined spaces, to secure casualties to stretchers with suitable lashings and to handle loaded stretchers with light ropes, as might be necessary. They needed anti-claustrophobia training and a high degree of skill in handling wounded persons in difficult situations. The doctors played a big part in their training, but otherwise they were given the same scope of training as the warden rescue teams.

In June, 1943, at the special request of Regional HQ, we staged a series of one week courses for rescue party leaders from other counties in the region. These candidates were billeted on the Leatherhead Urban District CD Rescue Service, arriving Sunday afternoon and leaving midday on the following Saturday. Many of these students were reluctant to attend the course, believing that they already knew everything that they needed to know. I attended at the depot every Sunday evening to give them a warm welcome and to prepare them for a tough week ahead. Notwithstanding my introduction, most of them got a shock when they arrived at the rescue school to discover that they were in the middle of an air-raid with fires burning and casualties trapped amid the wreckage and a few walking in need of immediate attention. By the time they had cleared up the situation to the best of their abilities, we were ready to start training in earnest. At the end of the course the students seemed reluctant to go home. Over four hundred leaders participated in this course.

At about this time the military were moving to regular courses week after week, so that we were running three different courses at the same time and the burden became even greater. It was, however, a tremendous challenge, and everyone felt that we were playing our part in the war.

15

Casualties Union moves into its second year

On 12 September, 1943, Casualties Union held its first annual reunion in the presence of the President W P Robinson, C.B.E., County Engineer and Head of the Surrey County CD Rescue Service, and Sir John Hodsoll, C.B., Director General of Civil Defence at the Home Office. Mr Robinson insisted that I took the chair. Both these gentlemen congratulated Casualties Union upon the high standard of its activities during the first year and the outstanding effect its services had had upon the quality of training at the rescue school. Several reports were received from such local districts as Egham, Caterham and Warlingham, Godalming and Hambledon. A number of leading members introduced new ideas they had been developing:

Bill Warn produced a non-spill reservoir for 'blood' so that one did not 'bleed' inadvertantly in the wrong place; rubber mask wounds; a sucking device for penetrating wounds of the chest.
John Stevens offered frothy 'blood' for 'bleeding' from the chest; burns faked with goldbeaterskin.
R Cortazzi produced a broken jaw device.

Backdrop for first reunion of Casualties Union, 1943.

W Steadman also produced a device to simulate a broken jaw.
Cyril Wallis offered a washable blood substitute.
Dr James Barford demonstrated the rocking stretcher method of artificial respiration.

At the end of the day Bill Warn and James Barford entertained us with some expert conjuring – a delightful relief after a year of highly concentrated endeavour.

But again I rush on, leaving important milestones unmarked. On 13 February, 1943, Wing-Commander Sir John Hodsoll, C.B., Director General of Civil Defence at the Home Office, was due to inspect the CD Services at Hawker Aircraft at Langley, near Slough and Casualties Union was invited to give special impact to these proceedings with a badly burnt pilot and a selection of other casualties. We were given a place to prepare in secret, somewhat remote from the inspection site, but it was arranged that we would be called in good time for the big moment. Somehow the arrangements went astray, because we never received that call. After it was all over the authorities remembered us in our remote stronghold and pleaded with us to stage another event with the burnt pilot outside the main entrance to the factory offices. As the pilot (Eric Buchler) lay there on a small grass patch by the entrance waiting for the first aiders to come to attend to his plight, who should come along but Sir Thomas Sopwith, who immediately started to attend to Eric and was still caring for him when the first aiders arrived. Such experiences indicate the standard of staging, behaviour and make-up that had already been achieved. Admittedly Sir Thomas had no idea that if he found a burnt pilot on that occasion, he might be a phoney, so he accepted him as genuine.

Of the Byfleet Ambulance members, Mrs Brenda Whiteley joined the staff of the rescue school fulltime, and unpaid, as leader of Casualties Union on New Year's Day, 1943 and Edward Akester joined us full-time in mid-March as Casualties Officer and instructor with special reference to wartime first aid.

Brenda became wardrobe mistress immediately, as well as prima donna amongst the 'casualties', and I have no hesitation in stating that, notwithstanding the wealth of talent that has been attracted to Casualties Union over the years, Brenda's performances for Dr Barford's lectures have never been surpassed. Modest and self-effacing as her character has always been, Brenda's leadership of the 'casualties' their final briefing as she put the finishing touches to make-up they had done themselves, was masterly. Despite the inevitable changes in personnel from week to week; the scenes at the 'pub', the 'cottage', etc. were remarkably consistent. Her mounting of the many special and differing demonstrations for 'red letter' occasions was equally masterful.

Edward Akester had already made a major contribution to the school's curriculum by his pioneer work in make-up of illness and injury. With the behavioural pattern instilled into the 'casualties', the practice of differential diagnosis, adequate for life saving and protective handling in wartime situations, became possible for the first time. Faced with such problems, always based on actual experience, which his skill produced, Edward found the application of dressings and bandages, while appropriate in principle as described in first aid manuals, were grossly inadequate to contend with the trauma of bombed casualties within blasted buildings. He set out, therefore, with Dr Barford's advice and encouragement to devise appropriate methods, using standard service issue dressings and bandages, to staunch bleeding in wounds inflicted to any square inch of the human body. He produced a magnificent chart, which showed in the simple language of diagram, how to apply every one of them. This work was subsequently published under the title *Practical Wound Treatment* by Edward Akester (Gale and Polden, Limited).

16

Enter the Military

On 30 March, 1943, Major Grange, Staff Officer Air Defence SE Army Command, visited the school to enquire about training facilities. On 19 April his deputy Capt. Morey, appeared and on 25 April the rescue school mounted its first training for military personnel. Such courses continued until the end of the war in Asia.

Colonel Taffe of the War Office visited the school on 7 January, 1944. Close on his heels Capt. Morey returned bringing Major E M Roe (First Canadian Army PAD Officer) and following up came a party of Canadian Officers introducing Sgt. Franey. Soon Canadian Officers, NCOs and servicemen were taking part in courses at the school, with Sgt. Franey caring for their regimental needs. From then onwards Canadian personnel took part in each course until the end of hostilities.

It was part of the arrangements with the military, that each Army should provide a representative on the rescue school staff for liaison purposes, and Sgt. Richardson and Sgt. Ritchie undertook these duties on behalf of the British and Canadian Armies respectively. The Armies also helped with military supplies necessary for the training; incendiary bombs, molotov cocktails and CAP canisters. It was an admirable arrangement and the two sergeants pulled their weight in full measure becoming assistant instructors in a short time.

Rescue workers searching and listening for casualties.

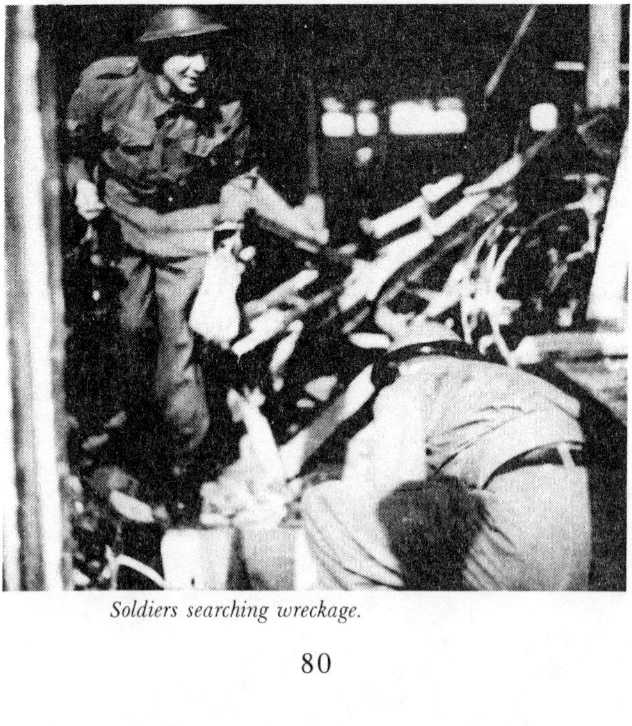

Soldiers searching wreckage.

Soldiers led by Instructor Arthur Knight, ready to lift casualty.

17

Second reunion of Casualties Union

Soon it was time to hold the second reunion of our casualties. It was decided to hold this before the winter because the indoor accommodation was so limited. It was, therefore, held outdoors on Sunday, 27 August. For this occasion Lanta Spurrier designed a CU Tree: with its roots in the debris of total war, the founder casualties, its trunk the CD Services and its branches, flowers and twigs. The branches were individual local authority groups, flowers were instructors and the twigs ordinary members, each by name. There were one or two saplings, growing by seeds wafted into adjoining areas, who had come to us for training. The whole issue was becoming widespread, so we also staged an exhibition of a number of exhibits, each directed towards making, in a very memorable manner, at least one pertinent point.

In this exhibition, we were proud to be able to pay tribute to the fighting services, which were sharing so greatly in the defence of Britain – RAF Fighter Command, AA and balloon barrage, which between them were intercepting so many of

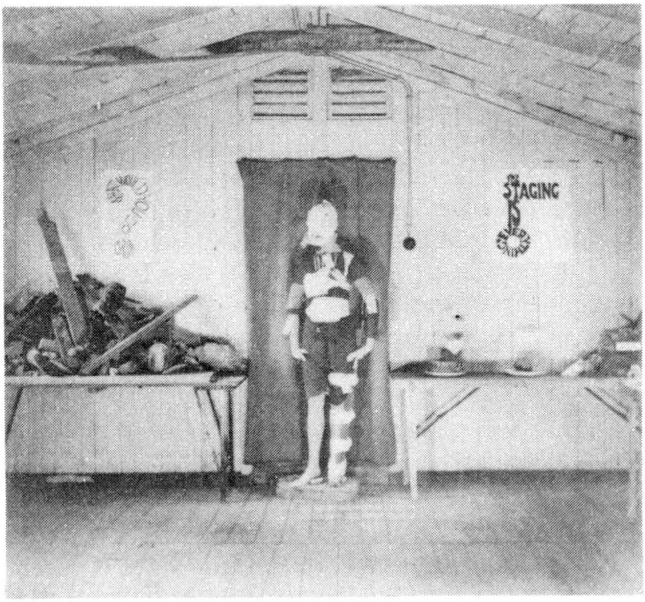

Three exhibits showing staging problems and a mannequin modelling many bandages.

Exhibit showing rope and some uses in rescue work.

the flying bombs (V1) and destroying them harmlessly over the open countryside. The intimidation value of the V1 was considerable. They were visible to the naked eye with a red flame being emitted from the rear. Its jet motor made a sadistic noise filled with menace, and it was obviously directed onto a large target – often London. However, when the engine cut out, the automatic steering changed so that many of them circled round a wide area and everyone within a mile of the point of cut-out felt threatened, until the devastating explosion at ground level was heard. The surface blast from one ton of high explosive demolished many dwellings and damaged large numbers of others over a radius of several hundred yards.

The V2, by contrast, was a rocket and arrived in utter silence to penetrate the ground before exploding. It caused more severe damage at the spot where it landed, but had little spread. If you heard the V2 explode you were safe; those near, who did not hear, were dead. Strategically, these bombs were designed for maximum morale destruction, and it must have been a great disappointment to the enemy that Londoners took them both in their stride – making jokes about their own problems and being encouraged by the futility of a cowardly enemy, who had not the guts to bring his destruction himself, but resorted to sending it by POST!

The enemy did the rescue school many good turns in his great and persistent folly. He bombed a large house adjoining the school at a time when our military intake was at its height, and obligingly enabled me to extend the training grounds materially. He bombed a factory, which made mannequin models for displaying women's glamorous clothing in the windows of fashion shops. The factory owner gave a number of these to the rescue school and they were invaluable for the display of sound bandaging, but were never used instead of human bodies for live or dead casualties. We feared that dummies would lead to carelessness, to bad habits in training, that could carry over into active service.

There was always a great temptation to clamber on top of the wreckage and pick one's way downwards, in spite of the obvious risks to survivors underneath. One exhibit entitled, 'Your boots, Sir', showed a cross section of wreckage within a glass cube. Head and hands of a victim could be seen amongst all the broken bits and plaster dust. On top were an inviting pair of service boots. Trained men never took their boots into such a dangerous position!

The difficulties of incident control which we were able to combat more effectively with the aid of our neighbours in Leatherhead and Dorking, as the enemy added to the size and scope of our training area.

We paid tribute to those who had already gained instructorship in Casualties Union. Pictured is the wartime certificate.

We displayed the Maewest and other life saving equipment available to pilots of aircraft in distress. We were determined to prepare the maximum number of people to help whenever an emergency occurred in their area.

Staging of situations for rescue training was shown in a triptych of exhibits; one showing unstable wreckage with a human foot just visible. The three were displayed under the heading, 'What would you do, chum?' – a cliché from the then very popular 'rag and bone man' sketches on the radio by Jack Warner.

Secondly, a careful choice of injuries was shown on a dummy complete with bandaging. It was entitled, 'In for a combined operation?' It was essential in choosing injuries for training purposes that there were real lessons to be learnt in debris management, choice of mode of handling and transport and the care needed to avoid binding in some of the wreckage with the injured limb.

Thirdly, there was a round of beef on a plate, a human foot on a plate and a badly mutilated hand amongst wreckage. Good make-up, and each of these exhibits was of the highest quality, was only of value if used in the proper context; the foot on the plate was revolting, while the round of beef was tantalisingly appetising; only the hand was challenging.

Perhaps I should explain about the human foot referred to in the previous paragraph. R Cortazzi, who was a dental mechanic, created an artificial foot out of dental jelly used for making moulds of the mouth when fashioning false teeth. First of all he made a plaster cast of someone's foot; then he made an appropriate skeletal form of cardboard, muslin and plaster; then he cast the jelly in between and finally he painted the skin all over, picking out the veins and adding toenails. It

THE CASUALTIES' UNION

INSTRUCTOR'S CERTIFICATE

OF C.D. SERVICES,
A MEMBER OF THE CASUALTIES UNION
HAVING ATTENDED A SPECIAL COURSE IN
CASUALTY MAKING, ACTING AND INCIDENT
PLANNING AND HAVING PASSED THE TEST
IS HEREBY APPOINTED A CASUALTIES'
UNION INSTRUCTOR.

SIGNED

W.P. Robinson
PRESIDENT

VICE PRESIDENT

ORGANISER

DISTINCTIONS

DATED

looked and felt human until one saw the absence of bloody flesh at the point of severance. Without any warning he sent this foot to me through the post. It did indeed feel like real flesh as I unwrapped it; every line and crack in the skin was there. The veins were a dull blue and the nails were perfect and suitably discoloured. It was some relief to discover that the skeletal form was hollow, being made of pure white plaster and gauze. That same foot, with bone and 'flesh' added, became something too real to be comfortable and it was made revolting by being placed upon a dinner plate with a dribble of blood.

Subsequently, Brenda and I visited Cortazzi and I assisted him in making a replica of Brenda's arm so that the rescuemen would have to contend with a woman's severed arm, as well as a woman whose arm had been torn off. It was an incredible replica and looked and felt absolutely real, but the extra thickness of flesh on the arm as compared with the foot presented difficulties in use. Cortazzi had overcome the problem of decay, but the jelly was tender and subject to cracking when handled. We had, therefore, continually to inflict more and more injuries to the arm, until it was entirely mutilated. Such was the pioneering spirit of early Casualties Union.

But to return to the exhibition; there was a display of the many and varied uses of ropes. There was an exhibit devoted to the many ways of rescue with the aid of a blanket, from slings to splints, which also gave a tip as to how to attach a rope to a blanket with the aid of a golf ball or large round pebble.

To enable the mass creation of 'casualties' on the military courses, as well as to ensure that all students learnt what the casualty went through during handling, we created a large series of mask wounds incorporated in a section of service uniform, that could be applied with appropriate coloured tape - coloured, I believe, with coffee, since we could only purchase white tape. Imagine someone lucky enough to have coffee in those days! This system saved the students time in changing clothing or damaging their uniforms. One exhibit was devoted to showing how the mask wounds were made and used. They covered everything from an arm severed at the shoulder, with the arm hanging in the sleeve by a few shreds; protruding intestines from a wound in the abdomen; to a sucking wound in the chest. As Brenda and her helpers fixed these wounds firmly to the soldiers' bodies over their uniforms, they briefed them how to behave with the injury they were supposed to have sustained. It was a brave effort and the immediate effect was good, because the clothing of the mask wound fitted naturally to the uniform the soldier was wearing.

This was a wonderful achievement by Brenda and her helpers, who included such pioneers as Cyril Wallis, Cortazzi, Roy Stokes, Bertie Lowe and Harry Davies. I hope that they appreciated the tribute that the display of their masks represented. One has, however, to admit that such injuries were not so easy to react to as those created upon one's own skin. Thus the second year of Casualties Union was well marked, and always as well as undertaking the essential business of electing their officers and reporting progress, the process of instructing all personnel was pursued in every way possible.

18

'Sambo'

Charlie Smallbone joined the staff on 1 September, 1943, to be a fulltime general factotum or orderly around the school. He was by trade a forester and his knowledge of trees was remarkable. He was a London-Brighton walker and in middle age still enjoyed making the trip. He worked for the Dorking and Horley Division of the County staff. Ill health suddenly made it necessary for him to work at a home base and a forester's duties were no longer appropriate. With the expansion of the rescue school to cater for military as well as civilians, I needed an extra hand. The Divisional Engineer phoned that this man was available and would I see him with a view to him filling my vacancy. I was happy to accept my colleague's recommendation.

When, however, the idea was put to Charlie, he made it clear that he could not commit himself until he had met and talked to me. In due course he came and interviewed me and luckily I managed to pass muster, for he accepted the job and turned out to be a treasure. Not only did his presence release a rescueman from orderly duties, but he took a pride in his work. He made it his special duty to care for the 'casualties' and I do not believe any 'casualty' came off the set without Charlie presenting him/her with a cup of fresh hot tea. His management of our slender resources of tea, milk and sugar was indeed masterly and he became a popular figure about the school. He also remembered to provide hot water for the ladies to ease the stern conditions.

Doc, who had a streak of naughtiness in his make-up, called Charlie 'Sambo' which I thought was a bit insulting, because he was a white man and he certainly was no slave, but Charlie did not seem to mind and for the rest of us too, he soon became Sam. He was a real pillar of the school. When I wrote my final report to the County Engineer, I was able to tell him that, 'Mr C Smallbone, in his capacity of porter and cleaner, has by his enterprise added materially to the comfort and well being of all who have been to the school. One and all, who have frequented the school, will have realised that to "ask Sam" was to know that the job was as good as done already. His service has been given in a truly delightful spirit'.

One other need was satisfied on the same day as Sam arrived. Lovesy and I spent far too much of our time away from training operations doing paper work and I needed a secretary, but I had no establishment for a clerical assistant, let alone a secretary. Speaking of my problems to the leader of the WVS (later to become WRVS) she clearly understood my need and my predicament because a few days later Mrs Noni Kinsbury, a member of WVS, 'phoned offering her services in that capacity. We met and thereafter Noni attended the rescue school several days each week. She was a great help to all concerned, saving Lovesy and me many hours in the office each week and thus improving the general efficiency. 1 September, 1943 was a good day!

19

In the front line

It is perhaps pertinent to record here some of the least comfortable situations to which we subjected our members, usually after finding actual problems on active service. Brenda told of her most painful experience as a 'casualty' – it was a situation in which she had been trapped by four pieces of debris. One lay across her shoulders; one lay across her lap; one lay behind her knees and the fourth was under her shins. She could not escape from her cramped position and all her weight was taken by her shins. It was near torture but she knew that, if only the one below her shins was removed, she could easily be removed seat first. When the rescuemen set to work to saw away the beam below her shins without first padding her shins, she was compelled to cry out in agony. Watching over her I realised that she was in great discomfort, but she was a magnificent actress, so I said to her, 'Please tell me when it ceases to be acting'.

'I was to ask a similar question many times in different situations, but this one impressed me greatly, so I determined to try it out myself. My admiration increased with every second it took the rescuemen to release me. The pain under the shin was bad and became excruciating while the sawing took place, but at least I had ragged trousers on: Brenda had only the remains of silk stockings! As a result of these tests, we alerted our rescuemen to the need for padding to such sensitive areas before sawing began, and our instructors to ensure that in such training this was always done. Also I made it a principle to try out all new situations myself before inflicting them on others. There were many salutary experiences.

However, I think that my most painful experience was being hauled out by my arms when my hips were wedged and I was acting unconscious. I was lying on a sloping floor so that I slipped tighter into the vee formed by the floor above which was collapsed on top of it. There were rafters under the upper floor, so the more the men pulled, the more I slipped into the vee, until I was totally wedged and I thought they would pull my arms out from my shoulder joints. If one man had climbed into the vee to hold me slightly up the slope, my hips would not have jammed.

As we moved into the second year of Casualties Union, it became increasingly obvious that instructor grade members were needed to maintain the standards in local areas. Therefore, on 23 April, 1944 a four day course, held on consecutive Sundays, was set up with the same instruction staff (Brenda, Doc, Akester and myself), as they had when they first joined. We had learnt a great deal in the intervening period, but none of us fancied ourselves as being in any way expert. We set out to polish up their individual performances in the several skills that are needed to create a convincing 'casualty', acting, staging, make-up and planning and to give them the additional skills in putting over training to would-be learners. At the latter some of us were by now pretty competent.

I remember addressing a batch of newcomers. Some of the candidates, who were very nervous, told me, 'It's different for you!' I do not think that it was really so different, but I understood what they meant. We are all nervous when making a speech, and it is important that we should be, because once we become confident there is a danger of becoming deadly dull. There were twenty-eight successful candidates and they were to form the basis of the senior instructor panel appointed a year later in June, 1945.

As I look back over the years I remember many heroic friends whose endurance as 'casualties' paved the way, I am certain, to the survival of other casualties in disasters from enemy action. There was Eric Buchler – an intrepid little man – whom we buried many times and placed in all the risky situations, which he accepted without demur and came back smiling for more. Later, when I learned that he had a silver plate in his skull following a First World War wound, I wondered what I had done to deserve such gallant help. There was a young man who had lost an arm above the elbow, who offered his deformity so that we could create a truly realistic severed arm – else we had to hide the well arm within the clothing and build up a stump on the shoulder. A middle aged woman, who had suffered a mastectomy, permitted us to build up a new 'breast' on her flattened chest, so that real practice could be gained in dressing and bandaging a wounded woman. There was warden Toledano, who had a heart condition and for those days a great age of seventy-nine when he joined us. He was a most faithful 'casualty' whom we dared not to lay down but who was of enormous value as one of the walking wounded and who could be trusted to tell his part of the story in a suitable manner, when questioned by the rescuemen. Robin Young, who suffered from a deplorable circulation and agonised in cold weather, faithfully undertook all the testing situations, providing she was allowed to wear an old fur coat. I mention just a few, but the reader can imagine the atmosphere at the rescue school which commanded the generous services of such people – all of them voluntary. They said, 'We are fighting the war too!' None of them was ever put off by gas, or fire, or the inevitable water, nor indeed by snow. They were the salt of the earth but so also were the rescuemen.

The rescue school was a dreadful, wonderful, glorious place, where people of all ages and from all walks of life gave of their best – not really surprising that the soldier students were impressed by this 'fighting to save life', as compared with their normal role of learning to kill. So many of the men at the end of the course tarried behind, loath to return to base and that was what so many of them told us, for somehow our little corner of hell was for them just a new pathway to heaven.

Towards the end of 1943 on 6 October, London Region held the final of their rescue competition at Tufnell Park and requested members of Casualties Union as 'casualties'. We were delighted and sent Cyril Wallis as one of the party. Cyril was to be severely injured and pinned down by an iron lamp-standard that had fallen in a raid. This suited Cyril very well and he prepared accordingly, but he became very angry when his briefing revealed that he was to wait until the competitors had lifted the standard off a 'dummy' and only then was he to take its place for first aid and removal. He asked whether the teams were all beginners; had they had any active service experience; whether they would be called out that night in the event of a heavy attack on London? When he received an affirmation to all three questions, he demanded that the dummy be removed and that he be put in its place at the outset. They surrendered and Cyril grew in stature as a result. Later he was to become one of Casualties Union's first instructors, and after the war was over he became a senior instructor and an honorary member – a talented and devoted member.

I recall that one cold November day the rescue party in training was exceedingly reluctant to locate and extricate me from the wreckage. Clothing was a big help at times to lift and move a casualty, but on this occasion I was wearing the minimum and the few items were well nigh worn out. Just a shirt and slacks held together by an old necktie knotted round my waist, socks and shoes; there I lay, prone, waiting to be released and getting colder every minute. At last after some half hour they discovered where I was. Finding the place where I lay rather restricted, they tried to draw me towards them by my clothes; my shirt rode round my shoulder, which added to their difficulty in dragging me, but they tugged a bit harder and, when eventually they got me clear, I was somewhat handicapped with my shoulders trussed up in my shirt and my trousers round my ankles. I was very cold, a bit sore where I had scraped along the concrete floor, and strangely clad in briefs and a tie round my waist, and a gold watch on my wrist. The latter was given me by my mother on my twenty-first birthday, which I wore throughout the war and have worn ever since.

Such experiences with dry lips, dust and grit on one's face, especially if one's condition precluded being given a drink, made us conscious of the tremendous relief, comfort and reassurance that having one's face wiped with a moist cloth could bring. We encouraged our men to carry such a moist cloth. To receive a drink from a service water-bottle is an experience never to be forgotten; the solid burp of water released from the bottle, as one lay on one's back, nearly choked one. We devised, therefore, a stopper with a feeding tube and an air-vent, so the casualty could be given a dribble of water that could be absorbed in a recumbent position. Therefore our water-bottles tended to have a pouch for a moist cloth and a place for the feeding stopper.

Comparing experiences, the 'casualties' were a great help towards forward progress. The 'casualty' can always be the best critic, and by saying 'You did this to me', we tried to give feedback to the services who handled us. The reader will recognise the very great importance of my being a 'casualty', so that I could plough back my experiences for the advantage of the rescue-first-aid-transport techniques.

During the war the accredited method of minimising shock was to keep the casualty warm and, unless there was serious bleeding, to give hot sweet tea to drink. 'Hot sweet tea', became a parrot cry. Such treatment is no longer considered the best thing to do, but in 1944-45, especially in that bitter winter, our ministering angel, Sambo, with his incredible service of hot sweet tea for every 'casualty' who came off the set, was a very great boon. We only had sweet tea because the tea went further when made in bulk and the sugar went further if it was also put into the mix. I recall that, in the years before the war I had never drunk tea with sugar; but when tea and sugar were short I stopped them making a special unsweetened brew for me, to help our rations to go as far as possible, and I too learned to drink, if not to like, hot sweet tea.

20

Further expansion of training

But the war continued and the threat to Britain got worse. More and more visitors came to the rescue school to weigh up what it had to offer, and to request its services. The Ministry of Home Security determined late in 1943 to set up a national school at Sutton Coldfield and for the first three weeks of 1944 the commandant designate was seconded to us at Leatherhead. We gave him free rein to attend any part of the school's activities that interested him, and he made the fullest use of our hospitality. When the Sutton Coldfield school opened, Akester, Harris and Dr Barford were seconded for one day each week and, at the beginning of June, Harris transferred full-time.

As the threat of invasion diminished, the Home Guard began to take an interest in rescue work and I gave my first talk to Home Guardsmen on 21 November, 1943, at The Bull PH, Leatherhead. Their training was on the lines of that given to the Warden Service and especially towards a supporting role for the CD Services, when they were hard pressed. This went on over an ever widening area of London Command. Next it was the turn of the Public Schools. Both Cranleigh and Christ's Hospital requested talks for their sixth-formers and a week later they came to Leatherhead for two days practical.

It was in the early months of 1944 that Dr Milicent M Pam and Doris Hammer conceived a new system of training using the new found realism which gave the opportunity of observing signs and symptoms of illness and injury that Casualties Union technique had made possible. They chose six members of Egham branch of Casualties Union and briefed them to enact a different disability – either illness or injury. They press-ganged six wardens to attend one each of the six victims. It was understood that each of these victims could be ques-

Soldiers removing casualty from snowbound site in Eindhoven.

Stretcher management over, through and past barriers.

tioned, examined and handled, without the application of any treatment other than perhaps postural, for one minute and a half. They would then make their diagnosis and pass on to the next victim. A time keeper created a rhythm of handling, diagnosis and movement to the next victim, until each of the wardens had examined all of the victims. Then the victims announced in turn the condition they were portraying and the signs and symptoms that might have been observed. From needing to press-gang six wardens on the Monday, there were queues of wardens eager to participate on the Friday. I was invited to attend a demonstration exercise held at Egham Fire Station where Paul Goudime, another of our members, assisted with the organisation. I was greatly impressed and to this day the diagnosis exercise has been a vital plank in routine training promoted by Casualties Union. The world owes a real debt of gratitude to Dr Pam, Doris Hammer and their Egham colleagues for this wonderful development in realistic training.

Every experience we had seemed to be ordained by God to prepare us for future service. When our second child was a few weeks old, my wife suffered severe whitlows on her fingers and could no longer bath our baby boy. So it fell to me to learn the art of handling a romping soapy babe in the bath. It was an exercise in which I delighted and I believe that I became thoroughly competent. Later I was to find that competence of great assistance when I found a large and very fleshy lady in a silk nightie trapped up to her waist in debris, after a flying bomb had brought her crashing down to the ground floor, complete with the roof and upper floor of her house. Hundreds of other houses were also damaged. It was not my own area so that, in spite of my uniform, I was just one of the helpers. She complained of pain in her back. Between the very podgy nature of her flesh and the silk nightie, she was as slippery as the soapy baby, but the same holds as held the baby secure helped me to support this large soft lady. The pain in her back was due to a splintered rafter against which she was leaning, fortunately without a wound, and this was eased immediately by a thickly folded blanket and the support I was able to give. Patiently two Home Guardsmen removed the debris, piece by piece, from her legs. She had no pain in her legs, but when finally both her legs were released we found

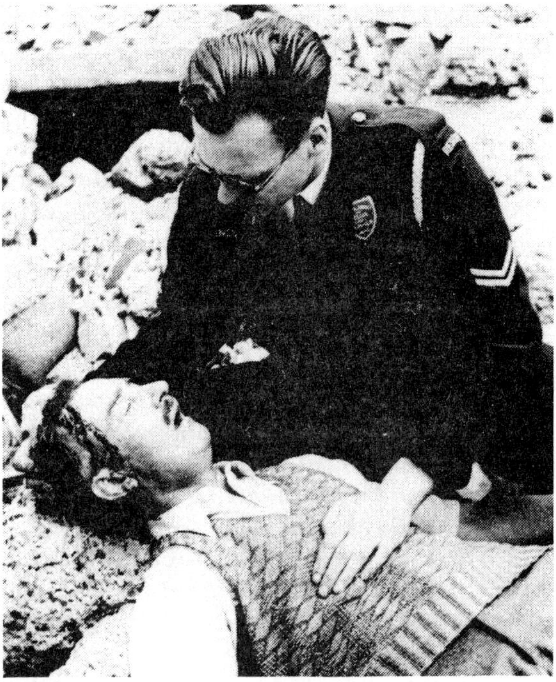

Roy Stokes, B.E.M. examines an unconscious casualty.

that the skin and flesh had been stripped away from both her legs from knee to ankle leaving the muscles completely exposed. The legs did not appear to be broken and I can only hope that she made a good recovery. One wished that one could follow a casualty to know how they fared and whether our help had been enough to minimise the damage. We continued to try our very best as experience grew and hopefully with it increased skills.

So much happened so quickly; in April the CD Staff College at Stoke d'Abernon made a routine of bringing their courses to Leatherhead to learn about battle training for Civil Defence. Groups of MOH, Controllers and Allied Nationals came, about forty at a time. Then came the National Fire Service. As a result, I made several visits to the Fire Service Staff College at Roedean, Brighton, to talk to their courses and a number of senior Fire Officers came on courses at Leatherhead.

It was the final round of the Regional Rescue Party Competition and I was to be one of the 'casualties' and in that capacity I was also to be one of the judges. I no longer remember the details of the setting, but I do remember why it all proved to be very exciting. The County Engineer, who was my chief, was present and everything went without a hitch for the first team. I was 'unconscious' and buried in the debris – all safely packed away but it looked dangerous. When the team had finished their tasks and I was released, my chief realised who the buried 'casualty' had been. He stormed into the preparation room, where we were getting ready for the second team, and without any preliminaries forbad me to be buried again – the risk was too great and he would not have me risking my life.

I took him outside and for the next half hour or so we walked up and down arguing. He was a man with whom nobody ever argued; it was therefore, I gathered later, considered by my colleagues to be both an unusual and an interesting occasion. Willy nilly the competition had to wait for me, for without me it did not scan. I contended that I knew precisely the amount of risk that I was taking; that, when he was not present, my judgement held sway; that in spite of the apparent dangers, as well as the real dangers we encountered, my rescue school had a clean record. Further that if he felt that he could no longer trust my judgement perhaps it would be better for me to be released to take up a commission in the Royal Engineers. In reply, I gathered that he did trust my judgement and was entirely satisfied with my handling of the Rescue School and the Rescue Service. It was because he thought so well of me, that he had been scared for my safety and desired to protect me from my own enthusiasm. I returned to my burial place amidst the debris and the contest was duly completed without further interruption.

21

Mark time two paces

This story reads as one of triumph after triumph, albeit with much hard work for every one of us. However there were disasters as well as joys and these in turn caused much concern. On 2 March, 1944, a bomber returning from a raid on the continent crash landed on the roof of my brother's bungalow in Norfolk, injuring his wife and three children. Fortunately the injuries were not serious but the bungalow was untenable. On 2 June, I got a spicule of debris in the pupil of my left eye. I called in at the chemist's shop and the chemist gently removed the foreign body with a soft brush. I blinked and all seemed well, though not surprisingly the eye was sore, but I soldiered on. After rest, whenever I awakened my eye was still very sore, but when I awakened on Monday, 5 June, my eye was unbearable, so I went to see my GP, who took one look at me and phoned his neighbour who was an eye specialist. I was quickly led from one consulting room to another in the next door house, where I received the biggest rousting of my life. The specialist said I was a wicked man, destroying my own health and risking my eyesight. However, he added that there was nothing wrong with my eyesight except for the scar, which healed while I slept and tore open every time I wakened to start work again. Gratuitously, he told me that I was exhausted. He then proceeded to bind up my face so tightly that I could not move my eye; he made me keep it so for a whole week and to lie on my back and rest completely for fourteen days. I went home and made arrangements to be away from duty for fourteen days and then got myself invited to a camp bed in a tent on a farm at Charwood, the home of Eric Buchler. I would travel down on the morrow, Tuesday.

When I awaked on the morrow, behold it was 'D' Day with all its risk of intensive bombing by the enemy. I phoned County Headquarters seeking instructions from the Deputy

County Engineer who, as Group Coordinating Officer for the engineering services of the ex-metropolitan Surrey with myself for his deputy, was my immediate boss. He said without any hesitation that I must go and get myself fit again; to take my fourteen days leave; that I would be back in time for the troubles to come. I did as I was told.

He was of course right. I arrived back on duty in time for the first flying bomb. I viewed the damage, caused by one that fell on Ewell on 21 June. On 27 July, one fell in the garden three doors from my own home in Chiltern Road, Sutton, Surrey. By the grace of God, the lady who lived in that house was the only casualty and she but slightly hurt. It happened about 6.20 pm – the very time when I would normally be returning home after the day at the rescue school to eat before going out on my evening assignment. However on this particular day, Brenda had needed to garage her car for maintenance and I had run her home, because there was no other direct form of transport. Her husband had insisted that I stayed for supper with them, because they had just received a food parcel from relatives in America. Lucky me! By the time I got home it was 9 pm. I found two Home Guardsmen in my house; they had cleared the broken glass from my bed and brushed the place through so I could get to bed. The back wall was spewed across the kitchen, the gas cooker stood at a cockeyed angle but still worked. All the outside doors were burst open and all the windows both upstairs and down were smashed, but the plumbing and sanitation still worked. The roof was largely stripped off but the Borough Council had spread a large tarpaulin over the lot making the house reasonably waterproof. Each of the hundred houses in the road were covered by tarpaulins. Many houses were also permanently open, so whoever of my neighbours who remained in residence – most were already evacuated including, thank God, my own family – awakened first and got a kettle boiled, took cups of tea to the others. Each of us had a free entry into the others' houses, even after they had been repaired, but only just until hostilities ended, when everything reverted to the status quo. It sounds incredible but it shows how truly difficult it is to learn the lessons of humanity and neighbourliness!

In November we were already looking forward to the end of hostilities, as far as Casualties Union was concerned, and we welcomed a visit from Dame Beryl Oliver, the British Red Cross Officer responsible for training. We gave her a demonstration of the uses of CU technique and its place in the pattern of training. She was highly appreciative and did us the honour of inviting Casualties Union to become part of the British Red Cross Society. In due course Mr W C Pape of St John Ambulance visited the school and extended to Casualties Union a similar invitation to become part of the St John organisation. Naturally, I discussed both these interesting invitations with my fellow members and finally declined, because we felt in honour bound to assist every organisation that could benefit from using our technique. One cannot help looking back and thinking in retrospect how much financial worry it would have saved us to have lived under one of their much bigger financial umbrellas. But, how then could we have been free to help the other, or the thousand and one other bodies we have served since? How could we have remained free to develop in many directions?

22

The British Second Army P.A.D. School

Then in the latter half of November, 1944, Sir John Hodsoll phoned me to say that he had received a request from Supreme Headquarters Allied Expeditionary Force (SHAEF) for me to be seconded to the British Second Army for service in Holland. Was I prepared to go? In less than one week I was flown from Northolt to Brussels in an RAF Dakota on my way to Eindhoven. We were a motley company on the plane, which had no seats; there was a General due to attend a meeting in Brussels; a number of servicemen of several arms; ten members of the NFS on their way to support the fire services in Belgium; and myself in my Civil Defence uniform and wearing the insignia of Commandant of the rescue school at Leatherhead. My rank looked similar to a Lieutenant Colonel, but in fact I had no standing and travelled on a civilian passport. We were held up for some time and circled the airfield, unable to land while dog fights took place below us. The General fumed and the flight commander appeared suddenly and gave us permission to smoke.

To my immense delight I was met at Brussels by my friend from the First Canadian Army, Teddy Roe, who was to be my military liaison. Teddy had been in Civil Defence in Harrow and was experienced in incident control before being called up. He was transferred to the Canadians, who needed someone to handle Passive Air Defence – the military title for clearing up the mess without striking back at the enemy. Teddy explained that the first students would arrive in three days time, but for the most part he could deal with that. On the fourth day a full course would arrive comprising personnel from the British Second, the Canadian First and the American Ninth Armies, with a few Dutch Civil Defenders for good public relations.

I enquired whether there were any soldiers nearby whom I

had trained in UK and he told me he had traced six Dutch Army Fire Service NCO's whom I had trained at Leatherhead, and that they would report for duty the following morning. They turned out to be a Godsend. I told Teddy that I should need a senior NCO or junior officer from each of the Armies to care for discipline within their own Army contingent. I enquired where I was to work and he told me they had earmarked a bombed-out clothing factory in the hope that this would be satisfactory. He also told me that the military authorities expected me to produce the same scope and quality of training I had provided for them in UK. He also promised to requisition any supplies or equipment that I needed.

I found myself attached to No.10 Garrison for messing and rations. Teddy Roe was attached to 8th Army Corps based at Roermond and then the fun began. Nobody knew what to make of me. Being attached to the British Second Army, I wore their flashes on both shoulders, but I was in a strange civilian uniform; however, my badges of rank resembled those of a Lieut. Col. and I was entitled 'Commandant of the British Second Army PAD School'. Sentries did not know whether to salute me or not. Civilian and Military Police each thought that I belonged to the other. It was a weird life for I was unique in Holland, but I had no difficulty with discipline or authority. What I said went without argument.

Teddy was billeted with the family Hazeu, father, mother and three children. Father, Hank, was a director of Phillips, the great electrical firm. I was billeted with the Castendijk family. The children were older and away from home and father was engineer in charge of the airfield and Controller of Civil Defence. All were charming people who did their utmost to make our stay in their country as comfortable as possible. They seemed pleased to have us, after the horror of the German occupation, and we needed their understanding and friendship.

I approved the site given to me for the school and soon, with the aid of those Dutch NCOs and local labour to create an office, lecture room, kitchen and messroom in the less damaged parts of the factory, we were ready to receive students. I was granted two medical orderlies to help with the preparation of some of the exercises, and two other orderlies to care for the cooking. When I discovered that my medical orderly corporal was a pastry cook in private life, it took me no time to rearrange their duties to everyone's satisfaction – I really do mean everyone. He was a splendid cook. As my three days of grace passed all too quickly, I gradually established where and what I would demand of my instructors and students. We used circular pipes in UK for anti-claustrophobia training, but the Dutch pipes were egg-shaped, which proved interesting. These had been located in the lee of a flank wall, which had escaped destruction during the previous bombing. Everything was ready to start the major course, and I had been ready in time to take my part in the short one-day course that Teddy had undertaken – so all was well. We had made a good start.

Imagine my chagrin as my students assembled the following day to discover that the flank wall had crumpled overnight crashing all over the first few exercises that were to follow my opening talk. A brief council of war led to my taking a little longer over my introductory lecture, amalgamating it with a second talk I had intended to deliver later in the day; meanwhile my splendid Dutch NCOs and local labourers restored the exercises perfectly, yet leaving debris littered about that looked well but did not interfere with our work. This was typical of the quality of co-operation I received at every turn. Every day each exercise was prepared and ready for me to set in motion.

But I still had a problem. I had to find 'casualties' and when I had recruited suitable volunteers, they would need to be trained. When I got back to my billet I asked my hosts to be good enough to invite the leading doctors in the city to meet me so that I could explain my needs and seek their help. They agreed readily and that same evening six doctors attended the Castendijk home. Having created a small split wound at the side of my forehead where a fall on a sharp edged granite kerb would have struck, and suitably damaged my knee, I slipped out into the rain and lay down in the wet muddy gutter, taking care that my face was smeared with mud as well as spattered with rain. I returned to the house and knocked. My host opened the door and ushered me into the only room with a light. Only one light was permissible in each occupied dwelling at that time. Of course I pleaded to be allowed to slip away to clean off some of the mud, but the doctors, seeing my bleeding head, sat me in a chair under the light. All was going according to plan. The light was on a pulley, so they pulled it down close to my head and had a good look. My host who was in the plot with me whispered in my ear that one of them was being sent to fetch sutures to stitch up my head. I begged him to tell them that it was all a fake, but they persisted that my head needed to be stitched. At that stage I slipped out of my chair, out of the room, gave myself a good wipe and a wash by the light of a torch and returned to explain why I needed their help and the sort of help I required. One of them was especially helpful to me. He was Dr Telema, but they all seemed to understand that this was an important feature of our training programme. Sixty-eight volunteers were found amongst the workforce at Phillips. Everybody was hungry and I was permitted to give a hot meal to each 'casualty', that I used during our training programme at the PAD school.

Dr Telema was present at many of the evening practice sessions held to rehearse what the casualties would need to know the following day. Everybody worked hard for those meals and it did my heart good to see them eat. It helped their families too, because if father had had a good meal there was more of the rations to share round the rest of the family. They deserved those meals, especially when the frost and snow came in the very bitter winter of 1944-45.

Whilst we are concerned with food, I must mention that I found it very difficult to cope with the switch from civilian to military rations. Presumably my stomach had got smaller because it was several weeks before I could enjoy my full rations. If we were going out for the evening at the Hazeu house, it was permissible for one to draw one's rations and it was a joy to see what our friends created when sharing rations with us. We were allowed a bottle of whisky each month and a number of bottles of beer. I found that some officers preferred beer to whisky and were happy to swap their whisky for my beer, which enabled me to share my whisky with our Dutch hosts. It seemed that most of the soldiers gave all their chocolate ration to the children.

After I had been there a few days the Hazeu family invited me as well as Teddy Roe to share their St Nicholastide festivities, which is very much a family affair in Holland and somehow we two fellows miraculously became part of their family. Sint Nikolas arrives at the door with black Peter carrying parcels for everyone. The head of the house goes to the door and receives these gifts, while the saint pursues his way to neighbouring houses. Each parcel is accompanied by a poem. This poem tells of the conduct of the recipient during the previous year and what amends must be made in the year then starting. There seemed to be an incredible number of souvenirs both for Teddy and for me – each with its own poem. As these were read out I began to understand some of the leading questions that had been asked of me on a previous visit. Doubtless Teddy had been questioned similarly about me. The talent for writing verse in English was remarkable and it was an unforgettable occasion. Each year when St Nicholastide comes round, I think back to that family and the warmth they extended to two friendly, but nevertheless foreign, servicemen waging war in their country, albeit to liberate them from a hateful enemy. I have kept in contact ever since.

Daily training continued without interruption except for a new set of students each week. The owner of the clothing factory re-made a number of service uniforms into the sizes and shapes needed for making mask wounds – it was super to have them made professionally and in use they were a great success. We had to use our initiative over make-up, because many of the materials used in UK were not available so close to the fighting line. I was able to get a small amount sent over from UK which happily was enough to tip the balance and a fair degree of realism was achieved.

We were granted the use of a bombed house about half a mile from the school and it was there that I staged an incident for the students to control and handle without further help. I went ahead to supervise the placing of the 'casualties'; while the contingent who were to deal with the situation were set to follow ten minutes later. Inevitably there was a gap in time. A middle aged Dutch lady riding on a heavy bicycle puffed up saying, 'For God's sake do something quickly, somebody is still moving'. I placated her and she waited to watch, as the action began. Then to our consternation a motorcade, or was it a convoy, came along and offered assistance, which we declined, but I was satisfied that the realism that I had sought for my students had been achieved.

Nevertheless I got a blistering message from command that I was not to use that site again as it made an interference with troops going up to the front. I believe that somebody who knew only scraps of the situation had over-reacted, but Command PAD Officer was adamant. My problem was that students became familiar with the school premises during the routine exercises and we had insufficient labour to keep changing things around, so that the final test could be in unfamiliar environment. 'C'est la guerre!' I believe our second best was adequate and we had to let it go at that.

Time was of the essence and we had to cram the maximum training into the seven days we were allowed. RAMC Officers joined me to give vital lectures on first aid in wartime. Sadly I only saw them for an hour - they came, they lectured, they were gone - so there was no chance to talk about tactics, our unique methods and our special problems. They remained strangely aloof - so unlike my wonderful panel of doctors at home who readily became part of the system. I wished I could have invited my Dutch medical friends but we had somehow to fit in with the military system.

The reaction of the soldiers was very similar to that of their colleagues in UK and I realised that the training overall was boosting the morale of the fighting services. They were learning to take care of themselves. Discipline was splendid and there developed a great rapport between staff and students on each course.

I only remember one disciplinary lapse; this occurred when an American top-sergeant refused to do a modest exercise involving fireman's lift. He vowed he would not do it because he might drop the other fellow and that would not be fair. I assured him that the other fellow did not mind. With his senior rank he could have picked the smallest man without anyone objecting, but refusal had brought him face to face with a top nobody - just a civilian in fancy dress who would not take 'no' for an answer, and eventually he carried me down a short ladder - and then back up again just to prove that he could! We parted the best of friends and he had a warm sense of achievement. Later on he carried me down a much longer ladder at his own request.

When Christmas came we had a brief respite and our pastry

cook did incredible things in a dustbin oven heated by a wood fire. Our administration officer, well advised by the cook, had requisitioned rations from the special Christmas list available, and no cake shop would have been ashamed to display the wares he produced. We invited our Dutch hosts and hostesses to a party in the school mess and neither they nor I had tasted such fare since I got involved in the Rescue Service in 1940. It was a delight to be able to acknowledge in this way with our warmest thanks, their hospitality, goodwill and friendship.

Our joys were quickly over and we were back to routine training. There came rumbles of a new enemy offensive and his adventure forcing the Ardennes salient suddenly moved into operation. I recall New Year's Day, 1945; I was standing on the rostrum addressing my new intake of troops. I had told them that I was aware that we might have some disturbances but that I proposed to continue as though all was normal. If they felt it was wise to take cover, I expected them to lie down on the floor, but nevertheless I should carry on. A quarter of an hour later there was a crump as bombs exploded not far away (on the local airfield, I learnt subsequently) and suddenly I was addressing an empty room – everyone, but myself, was on the floor. As I continued, my listeners reappeared, but we had a worrying time until the salient was closed and eliminated. Nobody could say that life was monotonous.

The weather was getting colder. Every morning there was ice in my washing water: sentries were reported to be dying at their posts out there in those dark cold nights. We all suffered for there was virtually no fuel except for the vehicles of war. I developed chilblains on both ankles from my heels to the calves of my legs. There were days when I could have crawled on hands and knees to breakfast in mess at No.10 Garrison Headquarters. The kindness of my hosts at the billet in making kettles of boiling water to unfreeze my chilblains warmed my heart, as much as the hot water eased and cured my legs. Nothing was too much trouble and indeed they had plenty of troubles of their own. No wonder that I learnt to respect and love the Dutch. Nevertheless, the training went on without interruption and the 'casualties' were heroes and heroines. As with their fellows in UK they endured whatever came to them without demur. They found a warm place in my heart and I was grateful indeed that I could give them a good hot sustaining meal – just the same as the soldiers. I began to eat well too.

One very cold afternoon, with frost and ice everywhere, a stranger rushed into my rescue school office and begged me to come to help with an accident in the street round the corner. It appeared later that he had been one of the volunteer 'casualties' from Phillips and therefore knew where to come. I hastened down the street and discovered that the casualty was a teenage girl who had fallen on the ice and struck her leg hard on the edge of the granite kerb. Her coat had given some protection, but her stocking was torn. By the time I reached the scene well-wishers had carried her into a bombed out shop where the sole remaining item of furniture was a chair. They had sat the girl on this chair, had crossed her knees with the injured left leg over the other, and horror upon horror, were in process of stripping off her stocking. When I arrived the child was crying out in agony and was clearly in great pain and distress. I could not speak any Dutch in those days and neither had she any English. Our communication therefore was solely by tone and touch.

She calmed when I spoke to her and as I gently placed my right hand under her left calf muscle and cradled her heel in the crook of my elbow, she ceased to cry and did not cry again before we handed her over to the hospital authorities. Having got her injured leg supported, I got one of the others to support her back, while another lifted up her right leg and I was able to tie her feet and ankles together and then her knees. By that time an ambulance arrived and we could place her on a stretcher. I went with her in the ambulance and she smiled sweetly as I handed her over at the hospital. I heard later that she had sustained a closed fracture of the left tibia and had made a good recovery. There is no doubt that the international language of first aid is TLC (tender loving care).

23

Broad Reconnaissance

I had agreed to go to SHAEF for two months and at the end of that time they sent over a relief from the Ministry of Home Security: regretfully not one of my people, but no doubt he did well. However, it was ordained that I should spend a week or so visiting the front line in many directions because the authorities had in mind, after a short break to check on things in the UK, to send me to the Italian front to undertake a similar mission.

At Maastricht we called on the US Ninth Army PAD Officer and while we were talking his phone rang. He answered then, covering the mouthpiece with his hand, he asked me whether I had ever seen an unexploded flying bomb. Noting my reply, 'No, but I would like to', he finished his call and in seconds we were on our way through the shattered remains of Aachen and ending up in an apple orchard. The US PAD Officer was a major; the officer in charge of the UXB a captain. He was delighted for the major and my escort Capt. Roe – somehow he had reverted to his sustantive rank – to see if they wished to kick the bomb, but it was as much as his life was worth to allow me to do so because I was a civilian. I pointed to the military flashes on my uniform. I told him that I had encountered more of these things at the receiving end than he had had hot dinners and that I had come a long way to kick such a one. 'They'd shoot me, if anything happened to you', he replied.

I doubted that I was that important when suddenly he changed his mind. He slipped his arm through mine and said, 'If anything happens to you, the same is sure gonna happen to me, so let's go and kick it together'. It was childish but kick it we did, arm in arm, and I have no doubt that we both felt better for doing so. I examined the brute. It was bigger than I had expected and rather crude and roughly finished. It had survived its mis-launch because it had planed along the top of the branches of the uniformly tall apple trees, and there it lay – smugly innocent – and I thought of the damage its fellows had wrought and the terror they had stricken into tens of thousands of hearts. I was strangely pleased to have shown my contempt.

Back in Maastricht we were fed on American rations which were subtly different from British. I recall sweetcorn and gum. We were taken to an officers' club where we were issued with a voucher to obtain a drink. The drink turned out to be peko. It was vaguely like discoloured gin – a bit strange; and I could imagine it being returned to store with a tag marked, 'This horse is unfit for duty'. I was content that the voucher only entitled me to one small glass.

We moved on to Nijmegen. The weather was bitter with frost and snow and the winterised Jeep gave little comfort to the traveller, except for the haybox in which the relief driver could warm his feet. When it was my turn as passenger I was indeed grateful to put my feet back into the box, hopefully to prevent the return of those chilblains. By common consent we changed drivers every half hour. We reached Nijmegen and saw the much-fought-over bridge across the Rhine and the battle-scarred remains of the eight-junction roundabout at the southern end. We proceeded as far as Elst. This was as far as we could go for the enemy was not far away in that snowy wasteland. Suddenly, almost under my feet, large guns opened fire. I nearly jumped out of my skin – I was only accustomed to the crump of bombs.

Arriving in Antwerp, as I got out of the Jeep to report to the police, a V2 rocket crumped nearby. The rule held; if one heard a V2 explosion, one was still all right.

Thence to Brussels where I found myself accommodated in the VIP hotel amongst generals, colonels and the like. They seemed happy to welcome such an odd looking individual and were interested to know what part I played. The next day I was back at the airport, thence to UK landing at Croydon airfield and so home again.

24

Return to Leatherhead

I was happy to be back at Leatherhead and to pick up the reins, ably held in my absence by Roland Lovesy and Brenda Whiteley. Italy dropped out of the war and the Germans were pushed back, so my commission to the Italian front never materialised. However my life interests had been widened by the experiences I had been given, and no doubt enriched too. People, not things, are clearly of prime importance in my life. The family were still evacuated so it was back to living alone again on civilian rations! The routine at the rescue school had become wholly military, about one quarter of the students being Canadian; this continued until the end of March, after which it was all British. The last course terminated at the end of May – just a few weeks after victory in Europe on 8 May, 1945.

I suppose the dullest assignment I had during the whole war was on behalf of SE Command in the famous Palm Court in Eastbourne, made famous in pre-war days by the celebrated violinist, Albert Sandler. There I met a vast number of officers destined to go to the Far East who were utterly bored with life in general and with their military assignment in particular. They certainly had no desire to listen to a civilian, even if he was talking about saving life instead of taking it. It seemed that they were all war weary and now that VE Day had come and gone they were to be sent a long, long way from home. They seemed happier when I had finished; but perhaps they were happy just because I had stopped! They were just waiting for VJ Day, but feared that they would become part of an occupying force.

25

Moves towards peacetime

On 24 April, Colonel E S Goss, M.C., assistant medical adviser to the Civil Defence Services invited me to his office in Whitehall where he introduced me to Dr White Knox, Surgeon-in-Chief to St John, and his deputy Dr M M Scott. Things moved quickly in those days. We were invited by Dr Scott to provide 'casualties' on 3 May for a Metropolitan Police First Aid Competition at McNaughton House, Euston. The casualties were Brenda Whiteley and Bill Warn. Everyone was very pleased and we were invited for another competition six weeks later. We have continued to assist the Metropolitan Police ever since.

We were also concerning ourselves with the future of the rescue school and of Casualties Union and our doctors pressed us that we should keep together to serve the needs of peace. If that was to be, we would need to lay a sound foundation for the future. We determined on a series of conferences involving some twenty-five leading members of the Union in mufti, to be held on Saturdays and Sundays for three weeks at the beginning of June, starting on the second and finishing on the seventeenth. We reviewed our own technique, organisation and administration: so much had woven round the facilities of the rescue school, with me as Commandant being somewhat of a dictator; with Lovesy, Brenda, Akester and Doc as authorities in their own right; whereas the others had achieved their positions by training and examination by these revered 'authorities'. In peacetime we would have no authority. So we pooled our knowledge; we invited doctors and other experts to talk to us about aspects of their work – medical, acting, staging, planning and later make-up.

Training in Acting

John Stevens recorded notes of the contribution of a teacher of dramatic art, Miss Frances Marsden, as follows:

1 The most valuable method of practice, especially in the early stages, is group acting. Students standing in a semicircle react at a given moment to a single emotion – fear, pity, horror, anxiety, curiosity, terror, etc. (A definite cause should be suggested). Or they may sit and register the shivering of fear (tense) and of cold (more relaxed), or practise crying, laughing and hysteria. In no case should the exercise be over-prolonged. Do not demonstrate BEFORE the practice, unless the students are hopelessly nervous. Individuals should have their faults checked, and may be encouraged to demonstrate and receive sincere group criticism.

2 More elaborate practices may take the form of exercises in mime. Here, however, it is better to begin with short scenes IN DIALOGUE, specially written for various situations and emotions: as these become familiar, the words can be reduced to a minimum and finally be discarded. At this stage, full attention may be devoted to expression and gesture – a valuable method for acting of blindness and deafness.

3 The mental aspect of acting is always the vital one. The student must think continuously, not in fits and starts; complete mental concentration should be practised, at first only for half a minute, but gradually lengthening the time. After acting fully, one is (and should be) absolutely exhausted. Before acting a part as a 'casualty', one should have a perfectly clear mental picture of one's thoughts at any given moment of the performance; vagueness is fatal to giving a clear impression. One should have a secondary emotion to fall back on in emergency, if (for example) one's partner is unexpectedly removed by the first-aiders. It is often possible (by fainting, collapse or temporary exhaustion) to give oneself a period of relaxation in a long act.

4 The body must work together as a whole: if a limb is not 'acting', it may spoil the whole effect. This should not,

however, lead to a wasteful tenseness: within reason as much of the body should be relaxed as possible, so as to have reserves for the 'big moments'.

5 The voice should also be practised. A single phrase can be spoken with various shades of emotion and character, eg. 'Where am I?' For different ages, 'Is he far away?' with fear, anxiety, tenderness, etc. 'I can't hear you,' would be suitable for decreasing the strength of the voice and 'Help!' for increase of pitch, volume and length, leading finally to self-control, or collapse, or a scream.

6 It is perfectly safe to fall, with the body relaxed, onto the flat of the hip, or via the knees onto the shoulder. A faint from a chair may go forward onto the arms, or sideways onto the shoulder.

7 Changes in medical condition should be practised in very slow motion, to ensure that the timing of the successive stages is convincing and accurate.

Medical thoughts about the future

Dr J E Haine, M.B., Ch.B., D.P.H., Medical Officer of Health for Guildford Rural District Council, who had assisted the rescue school as locum for Dr Barford on a number of occasions and had also become Medical Adviser to his district branch of Casualties Union, addressed the conference as follows:

'I am no seer, yet you have asked me to dip into the future. Never, perhaps, was it more difficult to see the trend of things to come. We are still too close to the tremendous happenings of the past few years.

'Education is exercising ever more and more influence upon every individual and is more appreciated than at any time in the past. Matters medical have caught the public imagination and are likely to remain as something of personal concern to everyone. The two factors are of fundamental importance to all of us here, as they affect our future. They have a vital influence on the future of Casualties Union. . . . I know the value of the pioneer work you have already done. I know its worth in the war effort and something of its potentialities for the future . . . The road will not be easy. It will demand your fullest effort – your constant watchfulness.

'There is no doubt whatever that the quality of training and the standard of first aid work increased and improved to a remarkable degree as a direct result of the assistance given by Casualties Union. The personnel, moreover, were much better fitted psychologically to deal with the real battle casualties. I recall also one incident – amusing to look back upon – where a whole exercise had to be cancelled because the ladies at the first aid point, highly efficient and thoroughly trained as they were, were completely knocked off their balance by the reality of your preparations and the realism of your casualties, and were quite unable to proceed. It was, moreover, very quickly apparent the best first-aid workers in the service were the members of your Union. Not that you had picked the best, but the training as a casualty gave a better understanding and knowledge, and a more realistic outlook than any form of training previously tried. Of that there is no shadow of doubt. . . .

'The future of first-aid training and the future of Casualties Union are closely related . . . Knowledge is important. The first-aid text book must have a place in your training. There is urgency for a thoroughly revised and up-to-date manual to replace the out-of-date books used by the two major first-aid organisations. A first-aid library is an asset to any first-aid organisation. I suggest that this should include books, pictures, X-rays, and up-to-date Press cuttings of items of first-aid interest from the medical Press.

'The British Red Cross Society and the St John Ambulance Brigade will, of necessity, continue to train large numbers of members. The police force, the National Fire Service, the nucleus of Civil Defence, the Armed Services, and the Merchant Navy Service will continue to require the adequate training of their members. There is quite a body of opinion that first-aid might be introduced into the school curriculum. Industry has developed an emergency service of first-aid that must be continued – most factories have their own aid posts and trained workers. In every walk of life a knowledge of first-aid is increasing in importance.

'Teaching must be one of your most important functions, the teaching of teachers. Have your principles carried far and wide until they reach every first-aid class in the country. As a teaching school or university you will have research in the forefront of your minds – research into the latest methods and advances. Scientific discoveries will be watched to see what application there may be in your own sphere. Again co-operation with medical science is all important if you are to advance. . . . Keep your independence. Independence coupled with virility of a youthful organisation will be amongst your greatest assets and allow you freedom to strike out in new directions. . . .

'The natural conservatism of our people is an obstacle to any new venture. . . . You must advertise yourselves. How many doctors know of your existence? . . . few . . . local divisions of the BMA arrange meetings for interest and instruction. I suggest that you could put on a turn that would not only interest them, but give them food for thought. Local secretaries of the BMA are the ones to interest. We are to have a national medical service of some sort in the next few years, and incidentally that includes re-organisation of hospital services, ambulance, nursing, and every branch of medical treatment. Make sure of your niche in the scheme of things.

'Let it be remembered that whatever you do will call for the very highest degree of efficiency. . . . The highest degree of skill, accuracy in detail – and may I add versatility and adaptability – must be your watchwords. You have shown no lack of enterprise and versatility in the past. You require those qualities even more to secure your future.

'Casualties Union is one of the really important and worthwhile things that have emerged from the war. Your experience must never be lost. It is, I believe, as yet only in its infancy. Practically all that I have suggested is no longer local or even county in character. Call it national, or call it universal. It is of universal application. Preserve your enthusiasm, your high tradition of service, prepare your organisation to meet with every development, and I have no doubts that you will take your rightful place in the scheme of things, and blaze a trail which all will regard with justifiable pride'.

What a challenge! Well at the end of those six days of discussion and thinking together, we felt that we knew a great deal about one another, both strengths and weaknesses. We agreed that we would all vote for senior instructor status by listing in order of merit, as judged by one's own personal opinion, the assembled company; the top ten would become the initial panel of senior instructors. The list chosen in order of votes was:

E C Claxton, Mrs B O Whiteley, J R Stevens,
Miss D I Hammer, A J Loarridge, Mrs O P Loarridge
F H Lowe, R Stokes, H Davies, Mrs L Spurrier

I believe that it was inevitable that, since the others had been able to accept Brenda and myself on their level following the conference, we should come out on top in their opinion, because it was obvious that we had been able to give most thought to all the problems in seeking solutions within the rescue school's routine. I believe that everyone was glad that we had met in mufti and had declared that, notwithstanding instructorship and senior instructorship, all members should be equal. We would be truly democratic. How near we have come to that ideal, I leave others to decide.

26

Third reunion of Casualties Union – end of the beginning

The final event at the rescue school was the third reunion of Casualties Union, when we wound up our wartime constitution and inaugurated a postwar Casualties Union. This took place on 1 July, 1945, and all those who had been members during the war and who elected to accept membership of the new Casualties Union became Founder Members. I was proud to be one of them. This great day was one of thanksgiving and renewed hope. There was a sadness at parting from old friends and a realisation that we had now undertaken a real challenge without any support from big brother, the rescue school, which was to be swept away on the following day, 2 July, 1945, when I was under instructions to start clearing the debris in readiness for returning the site to the convent. But the day was a day set aside for rejoicing. The members re-enacted in burlesque some of the highlights of their experiences through their years serving the school. There was George Evershed in the bed overhanging the basement steps and, as he struggled there, Bill Warn conjured eggs from his ankles, tomatoes from his knees and as he was rescued, Bill continued to 'find' all sorts of things to the absolute delight of the audience. Doris Hammer in her uniform as a VAD had us in stitches with Cyril Wallis in bed in an emergency ward. It seemed that there were others in the same bed – it was full of emergencies, but what emergencies!

The gang from Dorking area (A C Wallis, Miss R B Evans, R W Chilvers, Mrs L Spurrier, W C Songhurst and others) had a lot to convey in a song about smashing the swastika and they brought along a chart of the broken emblem to prove it. John Loarridge and other worthies kept pulling a chap through a pipe, but his head remained stuck where it was, while his legs became longer and longer. After that fellow had been well and truly extended to several times his normal length, they still

Third and Final reunion. Prominently displayed is the Casualties Union tree.

Bill Ward conjuring items from bedridden casualty, George Evershed – reminiscent of the initial day of Casualties Union.

Extricating long man from pipe. He stretched some 20 feet on recovery – too long for a stretcher, so they folded him.

continued pulling and a whole string of fellows came through like a string of sausages all tied together. Reg Tracy with his accordion led a party from Hambledon area into a series of wild songs; 'casualties' were handled lightheartedly and many of the rescue procedures were re-enacted with a difference – the difference each time created lots of laughs. Akester, another conjurer, did lots of things with Bill Warn, his casualty, who in turn did all sorts of things with Akester behind his back to the sheer delight of the assembly.

Before they departed one hundred and ten wartime members enrolled as founder members of the new Casualties Union to support all kinds of training, to minimise accident occurrence and to prevent suffering during the long awaited years of peace. The new Casualties Union existed with one hundred and ten members and would remain independent, free to assist anyone, indeed everyone who valued its help. Those who retired congratulated their colleagues and wished the venture every success.

It was an end but ALSO A NEW BEGINNING!

Successful co-operation between services – Rescueman carries Warden safely through 30 inch pipe.

APPENDIX

CASUALTIES UNION

ORIGINAL FOUNDER PRACTITIONERS

A list of those Rescue School Officers and 'Casualties' who pioneered the work at the rescue school and launched the inaugural day, 22.11.42.

Edward Akester, A.L. Andrews, James L. Barford, O. Elspeth Bristow, Eric Claxton, W.J. Maelor Evans, A. George Evershed, C.O. Hankin, Roland Lovesy, Brenda O. Whiteley.

FOUNDER OFFICERS OF CASUALTIES UNION

President:

W. P. Robinson, C.B.E., M.Inst.C.E., County Engineer

Vice Presidents:

Mrs B.O. Whiteley, Woking Ambulance Service,
Dr W.J. Maelor Evans, Asst. County Medical Officer,
Capt. A.M. Furber, County CD Training Officer.

Organiser:

E.C. Claxton, M.B.E. B.Sc., A.M.Inst.C.E., Commandant County Rescue School.

Deputy Organiser:

Capt. R. Lovesy, OC County Rescue School

Committee:

E. Akester, Woking Casualty Service, Miss N. Aronson, Farnham Casualty Service, Dr J.L. Barford, Medical Officer, County Rescue School, L.D, Doelberg, Caterham & Warlingham Rescue Service, A.G Evershed, Woking Wardens Service, W.J. Lancaster, Caterham & Warlingham Warden Service, R.C. Spener, Guildford Rural Warden Service, Mrs B.O. Whiteley, Woking Ambulance Service.

WARTIME MEMBERS WHO CONTINUED AFTER THE WAR TO BECOME THE CONSTITUTIONAL FOUNDER MEMBERS OF CASUALTIES UNION

C.J.M. Adie, W.H. Allaway, H.J. Allen, Miss B.K. Ashpital, J.M. Ball, Dr J.L. Barford, A. Barker, S.A, Beadle, Mrs D.A. Beadle, J.S. Beadle, Miss E.M. Brewer, R.R. Buckler, Miss N.M. Budgett, Mrs E.M. Bull, Miss N. Bullock, J.B. Bunce, Miss E.D. Callow, Miss A.O. Carr, Miss J.E. Challoner, A.H. Childs, R.W. Chilvers, E.C. Claxton, D.M. Collins, J.A. Connolly, Mrs D.R. Connolly, R. Cortazzi, Mrs N.E. Cowland, Mrs B.E. Crouch, F.R. Dahlstrom, H. Davies, W.V. Davies, F. Davies, Miss A Deadman, Mrs B.L. Dean, G.H. Dixon, Mrs J. Drew, Miss B.M. Edwards, Mrs P. Elshaw, Miss W.A. Elston, Miss R.B. Evans, Miss A. Fail, Rev.J.A. Fawcett, J.J.Q. Fox, S. Francis, Miss W. Franks, R.J. Gale, A.W. Garden, Miss O.F. Gibson, R.G. Giles, M.S. Godfrey, P. Goudime, Mrs W.M. Grey, Miss D.I. Hammer, C.H. Hammond, A.G. Harlow, G.B. Hartfree, Mrs M. Hassell, Miss J.M. Hayward, .A.T. Hearne, A.J. Heath, Cdr.W. Hill, Mrs G.K. Humphrey, H. Huntingford, R.A. Joye, S.R. Juniper, Mrs M. Jupp, N.B.G. Jupp, Mrs A.M. King, Miss B. King, E.E. King, F. King, Mrs E. Kinsbury, Mrs M. Knott, W.G. Kyte, A.J. Loarridge, Mrs O.P. Loarridge, F.H. Lowe, H.W. May, P.R. May, W.L. Milton, Mrs L.B. Mitchell, R.N. Mitchell, Miss M.D. Morley, L.A. Muddell, Mrs M. McCarthy, C.W. McCarthy, Miss N. Noakes, W.S. Occomore, Mrs J. Oddy, Miss J.N. Peachey, Miss S. Perrin, P.A. Perry, E.R. Pett, Miss K.E. Phelips, Miss I. Philpot, S. Picton, Miss B.J. Pilcher, Mrs W.E. Pilling, C.J. Pizzey, Mrs A. Playfoot, Mrs R.M. Price, H. de Quervain, A.J. Renmant, S.C. Rider, W. Riley, Miss H.M. Roberts, F. Rogers, A.G. Ryder, W.G. Sanders, P.G. Sargeant, Mrs W. Sayers, M.R. Saywell, W. Searle, A.F. Shaw, E.C. Sherwood, W.C. Songhurst, D.C. Speck, Mrs L. Spurrier, A. Stead, Mrs E. Stedman, W.T. Stedman, Mrs G. Stevens, J.R. Stevens, R.H.R. Stokes, R.S. Taplin, Mrs E. Thorman, W.C. Towndrow, Mrs M. Trye, Miss M.F. Tyte, D.J. Uridge, Mrs M. Wainwright, W. Wainwright, A.C. Wallis, Miss R.M. Ware, W.C. Warn, Mrs D.A. Webb, J.A. West, Mrs B.O. Whiteley, Mrs E. Winslow, F. Winsloe, D.A. Young, Mrs R. Young (later Grey).

WARTIME INSTRUCTORS

Mrs M. Bloomfield, Miss G.C. Braines, Miss A.O. Carr, Miss J.E. Challoner, Mrs N.E. Cowland, R. Cortazzi, H. Davies, Miss R.B.M. Evans, Rev J.A. Fawcett, Miss A.M. Freeman, Miss D.I. Hammer, A.J. Loarridge, Mrs O.P. Loarridge, F.H. Lowe, C.W. McCarthy, A.E. Patrick, J.C. Peddell, Miss W. Phillips, A.N. Prior, W. Riley, M.R. Saywell, Mrs L. Spurrier, W. Stedman, R.A. Tracy, A.C. Wallis, Miss G.L. Ware, Mrs E. Winsloe.

STAFF OF THE SURREY COUNTY RESCUE SCHOOL LEATHERHEAD

Commandant:
E.C. Claxton,
Officer-in-Charge:
Capt. R. Lovesy,
Medical Officer:
Dr J.L. Barford,
Instructors:
J. Bullen, S.H. Thomas, E. Akester, A. Harris, A. Knight,
O/C 'Casualties':
Mrs B.O. Whiteley,
Hon Secretary:
Mrs N. Kinsbury,
Caretaker:
C.T. Smallbone (Sambo).

THE SEVEN COUNTY RESCUE PARTIES AND THEIR LEADERS

Bagshot Division:
A. Knight, *F.S. Gearing,
Dorking & Horley Division I:
F. Peters,
Dorking and Horley Division II:
G. Crockford,

*Gearing took over the leadership when Knight became an instructor at the rescue school.

Guildford Division:
H. Snell,
Godstone Division I:
G. Longley,
Godstone Division II:
Dowall
Hambledon Division:
B.A. Riddle.